Python 强化学习实战：
使用 OpenAI Gym、TensorFlow 和 Keras

[美] 托威赫·贝索洛(Taweh Beysolow II) 著

敖富江 杜 静 张民垒 译

清华大学出版社

北 京

北京市版权局著作权合同登记号　图字：01-2020-3188

Applied Reinforcement Learning with Python: With OpenAI Gym, TensorFlow, and Keras
Taweh Beysolow II
ISBN: 978-1-4842-5126-3
Original English language edition published by Apress Media. Copyright © 2019 by Taweh Beysolow II. Simplified Chinese-Language edition copyright © 2021 by Tsinghua University Press. All rights reserved.

本书封面贴有清华大学出版社防伪标签，无标签者不得销售。
版权所有，侵权必究。举报：010-62782989，beiqinquan@tup.tsinghua.edu.cn。

图书在版编目(CIP)数据

　Python强化学习实战：使用OpenAI Gym、TensorFlow和Keras / (美) 托威赫·贝索洛(Taweh Beysolow) 著；敖富江，杜静，张民垒译. 北京：清华大学出版社，2021.1
　书名原文：Applied Reinforcement Learning with Python：With OpenAI Gym, TensorFlow, and Keras
　ISBN 978-7-302-57009-7

　Ⅰ. ①P… Ⅱ. ①托… ②敖… ③杜… ④张… Ⅲ. ①软件工具—程序设计 Ⅳ. ①TP311.5

中国版本图书馆CIP数据核字(2020)第238004号

责任编辑：王　军
封面设计：孔祥峰
版式设计：思创景点
责任校对：成凤进
责任印制：杨　艳

出版发行：清华大学出版社
　　　　　网　　址：http://www.tup.com.cn，http://www.wqbook.com
　　　　　地　　址：北京清华大学学研大厦A座　　邮　　编：100084
　　　　　社 总 机：010-62770175　　　　　　　　邮　　购：010-62786544
　　　　　投稿与读者服务：010-62776969，c-service@tup.tsinghua.edu.cn
　　　　　质 量 反 馈：010-62772015，zhiliang@tup.tsinghua.edu.cn
印　装　者：北京鑫海金澳胶印有限公司
经　　　销：全国新华书店
开　　　本：148mm×210mm　　　印　　张：4.5　　　字　　数：117千字
版　　　次：2021年2月第1版　　　印　　次：2021年2月第1次印刷
定　　　价：49.80元

————————————————————————————————————
产品编号：086653-01

谨以此书献给我的朋友和家人，在过去的十年，你们陪我走过了我人生的低谷。感谢你们让我成为最好的自己。如果没有你们，我就不能像现在这样快乐地生活。

译者序

近年来，AI 是最热门的技术之一，人们对"人工智能""机器学习""深度学习"和"强化学习"耳熟能详。其中，强化学习被视为最接近人类思维方式的机器学习范式，该类范式使得我们能够创建随时间演进的 AI。强化学习起源于 20 世纪 50—60 年代的最优控制，并在过去十年得到飞速发展。在经典游戏、投资管理、发电站控制、机器人模仿人类行走等领域取得成功，特别是在围棋智能程序 AlphaGo 中的成功应用，使其得到全世界的广泛关注。

本书介绍了一些强化学习算法理论，如策略梯度算法和 Q 学习算法等，并基于 Python 语言在不同经典游戏用例和做市策略用例中应用了强化学习算法。在学习过程中，读者还会接触到 TensorFlow、Keras 和 OpenAI Gym 等人工智能开发框架，并将学习如何在云资源中部署并训练强化学习解决方案。本书提到的应用案例有趣实用、实战性强，适合于对强化学习有初步了解的读者。通过学习，读者将能够研究、开发和部署强化学习解决方案。

本书主要由敖富江、杜静、张民垒翻译，参与本书翻译和校对工作的还有董德帅、林望群、晏杰、李华莹、田成平、沈洋、黄赪东、秦富童、李海莉、王震、李红梅、王跃华等。为精确表达原文含义，并使译文通顺、优美，译者们在翻译过程中查阅参考了大量中英文资料，并进行了广泛讨论。当然，限于译者自身水平与精力，翻译中的错误和不当之处在所难免，希望能够得到读者的积极反馈，以利于更正和改进。

希望读者阅读本书后能够进一步理解强化学习，并掌握基于 Python 语言的强化学习实现。祝愿读者能够从这里起步，成为强化学习领域的专家！

译　者

作者简介

Taweh Beysolow II 是一位数据科学家和作家,目前居住在美国。他拥有美国圣约翰大学的经济学学士学位和福特汉姆大学的应用统计学理学硕士学位。在成功退出与他人共同创立的初创公司后,他现在担任总部位于旧金山的私人股本公司 Industry Capital 的董事,在那里领导加密货币和区块链平台。

技术审校者简介

Santanu Pattanayak 目前就职于 GE Digital，是一名职业数据科学家，并且是深度学习书籍 *Pro Deep Learning with TensorFlow*(Apress，2017)的作者。他在数据分析/数据科学领域拥有 8 年的丰富经验，并具有开发和数据库技术背景。加入 GE 之前，Santanu 曾在 RBS、Capgemini 和 IBM 等公司工作。他毕业于加尔各答的 Jadavpur 大学电气工程专业，是一位狂热的数学爱好者。Santanu 目前正在海得拉巴的印度理工学院(IIT)攻读数据科学硕士学位。他还致力于数据科学黑客马拉松和 Kaggle 竞赛，且排名在全球 500 强之列。Santanu 在印度西孟加拉邦出生并长大，现在他和妻子一起居住在印度班加罗尔。

致　　谢

我要感谢 Santanu、Divya、Celestin 和 Rita。没有他们，本书不会取得成功。其次，我要感谢我的家人和朋友的一贯鼓励和支持。有了他们，生活变得丰富多彩。

前　　言

很荣幸第三次在 Apress 出版社出书！本书是我撰写过的最复杂书籍，但对于每一位数据科学家和工程师来说都物有所值。在过去的几年中，强化学习领域经历了重大变革，热爱人工智能的每个人都值得全身心投入。

作为人工智能研究的前沿，本书将是熟悉该领域状况以及最常用技术的绝佳起点。基于这一点，我希望读者能够从中汲取力量，从而继续自己的研究并在各自的领域进行创新。

目　　录

第1章　强化学习导论 ·· 1
　1.1　强化学习的发展史 ·· 2
　1.2　MDP 及其与强化学习的关系 ··· 3
　1.3　强化学习算法和强化学习框架 ··· 5
　1.4　Q 学习 ·· 8
　1.5　强化学习的应用 ··· 9
　　　1.5.1　经典控制问题 ··· 9
　　　1.5.2　《超级马里奥兄弟》游戏 ··· 10
　　　1.5.3　《毁灭战士》游戏 ·· 11
　　　1.5.4　基于强化学习的做市策略 ·· 12
　　　1.5.5　《刺猬索尼克》游戏 ··· 12
　1.6　本章小结 ··· 13

第2章　强化学习算法 ·· 15
　2.1　OpenAI Gym ·· 15
　2.2　基于策略的学习 ··· 16
　2.3　策略梯度的数学解释 ··· 17
　2.4　基于梯度上升的策略优化 ·· 19
　2.5　使用普通策略梯度法求解车杆问题 ··································· 20
　2.6　什么是折扣奖励，为什么要使用它们 ································ 23
　2.7　策略梯度的不足 ··· 28

2.8	近端策略优化(PPO)和 Actor-Critic 模型		29
2.9	实现 PPO 并求解《超级马里奥兄弟》		30
	2.9.1	《超级马里奥兄弟》概述	30
	2.9.2	安装环境软件包	31
	2.9.3	资源库中的代码结构	32
	2.9.4	模型架构	32
2.10	应对难度更大的强化学习挑战		37
2.11	容器化强化学习实验		39
2.12	实验结果		41
2.13	本章小结		41

第 3 章 强化学习算法：Q 学习及其变种 … 43

3.1	Q 学习	43
3.2	时序差分(TD)学习	45
3.3	epsilon-greedy 算法	46
3.4	利用 Q 学习求解冰湖问题	47
3.5	深度 Q 学习	50
3.6	利用深度 Q 学习玩《毁灭战士》游戏	51
3.7	训练与性能	56
3.8	深度 Q 学习的局限性	57
3.9	双 Q 学习和双深度 Q 网络	58
3.10	本章小结	59

第 4 章 基于强化学习的做市策略 … 61

4.1	什么是做市	61
4.2	Trading Gym	63
4.3	为什么强化学习适用于做市	64
4.4	使用 Trading Gym 合成订单簿数据	66
4.5	使用 Trading Gym 生成订单簿数据	67

4.6 实验设计 ··· 68
    4.6.1 强化学习方法 1：策略梯度 ······························· 71
    4.6.2 强化学习方法 2：深度 Q 网络 ··························· 71
4.7 结果和讨论 ·· 73
4.8 本章小结 ··· 74

第 5 章 自定义 OpenAI 强化学习环境 ································· 75
5.1 《刺猬索尼克》游戏概述 ··· 75
5.2 下载该游戏 ·· 76
5.3 编写该环境的代码 ·· 78
5.4 A3C Actor-Critic ··· 82
5.5 本章小结 ··· 88

附录 A 源代码 ·· 91

第 1 章
强化学习导论

阅读过我以前的著作——*Introduction to Deep Learning Using R*(Apress，2018)和 *Applied Natural Learning Using Python* (Apress，2017)的读者，很荣幸你们能够再次成为我的读者。新读者，欢迎你们！在过去的几年里，深度学习软件包和相关技术持续不断地涌现和发展，推动了许多行业发生革命性的变化。毫无疑问，强化学习(Reinforcement Learning, RL)是该领域最引人注目的篇章之一。本质上，强化学习是众多 AI 技术的应用，例如能够学习如何玩电子游戏或下棋的软件。强化学习的优点在于，假设问题可以建模为包含动作、环境、agent 的框架，则 agent 可以自行通晓各种任务。问题可以覆盖从破解简单的游戏到更复杂的 3D 游戏，再到教授无人驾驶汽车如何在不同的地点接送乘客，以及教授机械臂如何抓取物品并将它们放在厨房柜台之上。

训练良好并部署的强化学习算法意义显著，因为它们追求将人工智能技术应用到更广阔的领域，而不只是我之前著作中提到的一些狭义的 AI 应用场景。算法不再只是简单地预测一个目标或标记，而是操纵环境中的 agent，这些 agent 具有一组动作集，可选择其中的动作以实现目标/奖励。一些公司和机构投入大量的时间研究强化学习技术，如 Deep Mind 和 OpenAI，它们在该领域取得了一些领先的解决方案突破。首先，让我们简要概述该领域的发展史。

1.1 强化学习的发展史

在某种意义上讲，强化学习是最优控制(optimal control)的重塑，而最优控制是从控制理论扩展而来的概念。最优控制起源于 20 世纪 50—60 年代，当时它被用来描述一类问题，即人们试图达到某种"最佳"标准以及达到这一标准需要的"控制"法则。通常，将最优控制定义为一组微分方程。这些方程定义了一条使得误差函数最小化的路径。最优控制的核心理论由理查德·贝尔曼(Richard Bellman)所做的工作推向顶峰，尤其是他提出的动态规划(dynamic programming)方法。动态规划是 20 世纪 50 年代提出的一种优化方法，强调通过将大问题分解为更小且更易于解决的小问题来求解。动态规划被认为是解决随机最优控制问题的唯一可行方法，更一般意义上，最优控制方法通常都属于强化学习范畴。

贝尔曼对最优控制的最著名贡献是哈密尔顿-雅可比-贝尔曼方程(Hamilton-Jacobi-Bellman equation，HJB 方程)。HJB 方程如下：

$$\dot{V}(x,t) + \min_{u}\{\nabla V(x,t) \cdot F(x,u) + C(x,u)\} = 0,$$
$$\text{使其满足 } V(x,T) = D(X)$$

其中 $\dot{V}(x,t)$ 是 V 关于时间变量 t 的偏导数。"·"表示点乘，$\dot{V}(x,t)$ 为贝尔曼价值函数(未知标量)，代表系统从时间 t 处的状态 x 开始，按照最优方式控制，直到时间 T 时的成本，C 为标量成本速率函数，D 为最终效用状态函数，$x(t)$ 为系统状态向量，$x(0)$ 假设已知，其中 $0 \leqslant t \leqslant T$。

该方程得出的解是价值函数或者是给定动态系统的最小成本。

HJB 方程是解决最优控制问题的经典方法。此外，动态规划通常是解决随机最优控制问题的唯一可行方式或方法。马尔可夫决策过程(Markov Decision Process，MDP)是此类问题之一，研究动态规划方法正是为了帮助求解它们。

1.2 MDP 及其与强化学习的关系

我们将 MDP 描述为离散时间随机控制过程。具体来说，将离散时间随机过程定义为下标变量是一组离散或特殊的值(相对于连续值来说)的随机过程。MDP 特别适用于结果受过程中参与者影响且过程也表现出一定程度随机性的应用环境。MDP 和动态规划因此成为强化学习理论的基础。

简而言之，基于马尔可夫特性，我们假设未来只与现在有关，而与过去无关。此外，如果一个状态对未来的描述与我们具有的所有历史信息相同时，则认为该状态是充分的。从本质上讲，这意味着当前状态是唯一与之相关的信息，不再需要所有历史信息。从数学上讲，一个状态具有马尔可夫特性，当且仅当

$$P[S_t+1|S_t] = P[S_t+1|S_1,...,S_t]$$

马尔可夫过程本身被认为是无记忆的，因为它们是状态间的随机转移。进一步，我们认为它们是状态空间 S 上的元组 (S, P)，其中状态通过转移函数 P 进行变更，转移函数定义如下。

$$P_{ss'} = \mathbb{P}[S_{t+1} = s' | S_t = s],$$

其中 s 表示马尔可夫状态，S_t 表示下一状态。

该转移函数描述了一种概率分布，其中分布是 agent 可转移到的所有可能状态。最后，从一种状态转移为另一种状态时会获得奖励，在数学上将其定义如下。

$$R_s = \mathbb{E}[R_{t+1} | S_t = s],$$
$$G_t = R_{t+1} + \gamma R_{t+2} + \gamma^2 R_{t+3} + \cdots + \gamma^{k-1} R_{t+k}$$

其中 γ 表示折扣因子，$\gamma \in [0, 1]$，G_t 表示总折扣奖励，R 表示奖励函数。

因此，马尔可夫奖励过程(MRP)元组定义为(S, P, R, γ)。

基于上述描述的公式，图 1-1 所示的示例可视化展示了马尔可夫决策过程。

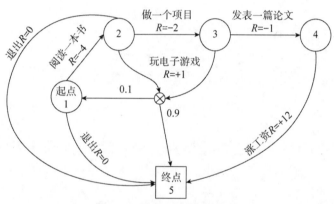

图 1-1　马尔可夫决策过程

图 1-1 显示了 agent 如何以不同的概率从一种状态转移到另一种状态从而获得奖励。最佳情况是，在给定环境参数下，我们如何在给定周期内成功学习累积了最高奖励的过程。从本质上讲，这是强化学习非常朴素的解释。

试错学习(trial and error learning)是推动强化学习发展的另一类重要技术，它是研究动物行为的一种方法。具体来说，已证明它对于理解基本奖励和"增强"不同行为的奖惩机制很有意义。但是，"强化学习"一词直到 20 世纪 60 年代才出现。在此期间，特别是马文·闵斯基(Marvin Minsky)引入了"信用分配问题"(Credit-Assignment Problem，CAP)的思想。闵斯基是一位认知科学家，他一生的大部分时间致力于人工智能研究，如他的著作 *Perceptrons* (1969 年)和描述信用分配问题的论文"Steps Toward Artificial Intelligence"（1961年）。针对为获得成功做出的所有决策，CAP 求解对应的信用分配方式。具体而言，许多强化学习算法直接致力于求解该问题。虽然如此，试错学习并不是很流行，因为神经网络方法(以及一般的监督学习)吸引了 AI 领域的大部分研究兴趣，例如 Bernard Widrow 和 Ted Hoff 提出的创新性研究。然而，在 20 世纪 80 年代，随着时序差分(Temporal Difference，TD)学习的兴起以及 Q 学习的发展，人们对该领域的研究兴趣再次兴起。

具有讽刺意味的是，明斯基着重指出，TD 学习受到动物心理学另一个方面思想的影响。它源自两类刺激，即一种主要增强物与一种次要增强物(译者注：任何能够加强行为发生的刺激都可被称为增强物)，它们成对出现，从而影响行为。但是，TD 学习本身主要由 Richard S. Sutton 提出。因为他的博士论文介绍了时间信用分配思想，他被认为是强化学习领域最具影响力的人物之一。该思想指出奖励可被延迟，特别是在细粒度的"状态-动作"空间中。例如，赢得一盘国际象棋需要采取很多步动作，才能获得赢棋"奖励"。这样，奖励信号对时间上离得较远的状态没有显著影响。因此，时间信用分配解决了对这些细粒度动作的奖励方式，其对时间上遥远的状态具有较有意义的影响。Q 学习以产生奖励的"Q"函数命名，它有一些创新，并聚焦于有限马尔可夫决策过程。

Q 学习的兴起，开拓了强化学习的新局面，使其得到持续改进，并代表了 AI 的前沿技术。完成本概述之后，接下来更具体地讨论读者将学习的内容。

1.3　强化学习算法和强化学习框架

强化学习与传统机器学习中的监督学习领域相似度很大，尽管存在一些关键差异。在监督学习中，存在一个客观目标，即我们训练模型的目的是基于给定观察值的输入特征进行正确的预测，而预测结果可以是类标签或特定的值。特征类似于环境中给定状态的向量，我们通常将其作为一系列的状态合集输入给强化学习算法，或一个状态接着一个状态地分别输入。但是，它们的主要区别在于，对于特定问题，强化学习没有必要总是只给出一个"答案"，因为强化学习算法可以通过多种方式成功求解一个问题。因此，我们显然希望所选择的答案能够尽快且尽可能有效地求解问题。这正是我们选择模型的关键所在。

在前面概述强化学习历史的过程中，介绍了一些定理，在后几章中将详细介绍这些定理。但是，由于本书是"实战"型书籍，因此在介绍理论时还必须同时提供一些示例。在本书中将占用大量篇幅讨论强化学习框架 OpenAI Gym，及其如何与其他深度学习框架进行交互。OpenAI Gym 是一种便于部署、比较和测试强化学习算法的框架。而且它具有很大程度的灵活性，使得用户可以在 OpenAI Gym 中应用一些深度学习方法，本书后续在证明各种概念的过程中将会利用该功能。下面给出了一些简单的示例代码，这些代码利用了 gym 软件包和绘图功能，其中绘图功能用于显示训练过程中产生的图像(见图 1-2)。

图 1-2　车杆游戏

```
import gym

def cartpole():
    environment = gym.make('CartPole-v1')
    environment.reset()
    for _ in range(50):
        environment.render()
        action = environment.action_space.sample()
        observation, reward, done, info = environment.
```

```
step(action)
print("Step {}:".format(_))
print("action: {}".format(action))
print("observation: {}".format(observation))
print("reward: {}".format(reward))
print("done: {}".format(done))
print("info: {}".format(info))
```

查看该代码,注意到在使用 gym 时,必须初始化算法所处的环境。尽管通常使用该软件包提供的环境,但也可以为定制任务(例如 gym 中未提供的电子游戏)创建自己的环境。接着,让我们继续讨论代码中定义的其他变量,这些变量的终端输出如下所示。

```
action: 1
observation: [-0.02488139 0.00808876 0.0432061 0.02440099]
reward: 1.0
done: False
info: {}
```

这些变量的含义如下。

- action——环境中 agent 所采取的动作,随后将产生一个奖励。
- reward——赋予 agent 的奖励,暗示了对应于完成某目标的动作质量。
- observation——由动作产生,指执行某个动作后的环境状态。
- done——布尔类型,指示是否需要重置环境。
- info——字典类型,存储各类调试信息。

这些动作的处理流程如图 1-3 所示。

图 1-3　强化学习算法和环境的处理流程

为了提供关于车杆(Cart Pole)游戏的更多信息，图1-2显示了关于小车和杆子的电子游戏，其目的是成功平衡小车和杆子，使杆子永不倒下。因此，相应的目标是训练某种深度学习或机器学习算法，以完成该任务。我们将在后续章节中解决这个特定问题。本节的目的只是简要介绍OpenAI Gym。

1.4 Q学习

在前述内容中简要讨论了Q学习。但是，有必要强调一下，本节是本章的重要部分，本节将讨论Q学习内容。Q学习的特征在于存在一些"警察"，"警察"会通知agent在不同情况下应采取的动作。虽然它不需要模型，但我们可以使用一个模型，模型通常采用有限马尔可夫决策过程。具体来说，本书将介绍Q学习的各类变体，包括Q学习、深度Q学习(DQL)和双Q学习(见图1-4)。

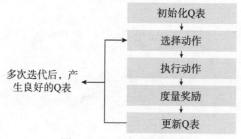

图1-4　Q学习流程图

本书将在专门介绍这些技术的章节中对它们进行更深入的讨论。但是，鉴于问题的复杂性，Q学习和深度Q学习各有其优点，虽然两者通常都存在类似的不足之处。

Actor-Critic 模型

本书介绍的最高级模型是 Actor-Critic 模型，它由 A2C 和 A3C

两类模型组成。二者分别表示 Advantage Actor-Critic 模型(优势 Actor-Critic 模型)和 Asynchronous Advantage Actor-Critic 模型(异步优势 Actor-Critic 模型)。虽然两者实际上是类似的,但不同之处在于后者具有多个相互配合工作的子模型,并且这些子模型独立更新参数,而前者同时更新所有子模型的参数。这些子模型更新参数的粒度更细(动作到动作粒度),而不是像许多其他强化学习算法那样以回合为粒度进行更新。图 1-5 为 Actor-Critic 模型的可视化示例。

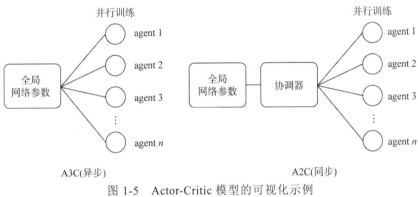

图 1-5　Actor-Critic 模型的可视化示例

1.5 强化学习的应用

在向读者全面介绍强化学习概念之后,我们将开始处理多个问题,重点是向读者展示如何在云环境中部署解决方案,我们将在云环境中进行强化学习算法训练和应用。

1.5.1 经典控制问题

由于最优控制领域已经拥有大约 60 年的历史,因此存在大量经典问题,我们将优先处理这些问题,读者在其他强化学习文献中也会经常看到它们。其中之一是车杆问题,如图 1-2 所示,该游戏要求用户使用最佳选项集来尝试平衡车杆。图 1-6 中显示了另一个问题，称为《冰湖》游戏。在该游戏中，agent 学习如何穿越冻结的湖面而不踩在会导致其落水的冰上。

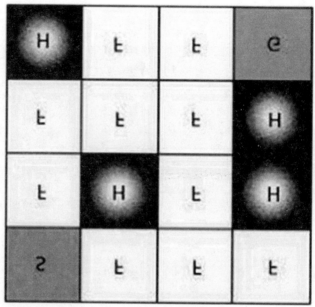

图 1-6 《冰湖》游戏

1.5.2 《超级马里奥兄弟》游戏

迄今为止，《超级马里奥兄弟》是最令人喜爱的电子游戏之一，它被证明是展示将人工智能中的强化学习技术应用于虚拟世界的最佳方式。借助 py_nes 库，可以模拟《超级马里奥兄弟》游戏(见图 1-7)，然后利用游戏中的数据来训练模型自行通关。我们将只专注于一个级别的关卡，并在这个应用中使用 AWS(亚马逊云服务)资源，使读者有机会获得该任务中的经验。

图 1-7 《超级马里奥兄弟》游戏

1.5.3 《毁灭战士》游戏

此处介绍的经典强化学习应用示例是如何学习玩简单级别的《毁灭战士》(Doom)游戏(如图 1-8 所示)。这款游戏最初发布于 20 世纪 90 年代,运行于 PC 上,游戏的规则是在通过整个关卡的过程中成功杀死玩家面对的所有怪物或敌人。在给定动作范围、可用的软件包,及其他有帮助的属性后,可以完美应用深度 Q 学习算法。

图 1-8 《毁灭战士》游戏截图

1.5.4 基于强化学习的做市策略

不同的自有资金交易公司的共同策略是通过向参与者提供流动性(目的是以任何给定的价格买卖资产)来赚钱。尽管该策略已具有成熟技术,但由于目标相对简单且该领域数据丰富,因此这是应用强化学习的绝佳领域。我们将使用 Lobster 的限价订单簿(limit order book)数据,该网站包含大量用于此类实验的完美订单数据。在图 1-9 中,可以看到一个订单簿的示例。

SPDR S&P 500 ETF TR TR UNIT

Orders Accepted: 1,153,586
Total Volume: 7,689,062

	TOP OF BOOK		LAST 10 TRADES		
	SHARES	PRICE	TIME	PRICE	SHARES
ASKS ↑	11,000	180.07	14:42:13	180.03	100
	12,500	180.06	14:42:11	180.02	100
	12,900	180.05	14:42:11	180.01	100
	9,700	180.04	14:42:09	180.01	100
	1,100	**180.03**	14:42:09	180.01	200
BIDS ↓	6,400	**180.02**	14:42:08	180.01	100
	9,700	180.01	14:42:06	180.01	100
	9,600	180.00	14:42:06	180.01	100
	14,700	179.99	14:42:06	180.01	100
	11,500	179.98	14:42:06	180.01	100

图 1-9 限价订单簿

1.5.5 《刺猬索尼克》游戏

另一款适合应用不同模型的经典游戏是《刺猬索尼克》,如图 1-10 所示。除本章之外,我们将引导读者从零开始逐步创建自己的环境,使读者能够封装一个应用 OpenAI Gym 和自定义软件的环境,然后训练自己的强化学习算法以解决关卡问题。这里将再次利用 AWS

资源进行训练,并借用其他电子游戏示例(尤其是《超级马里奥兄弟》游戏)中使用的相同过程。

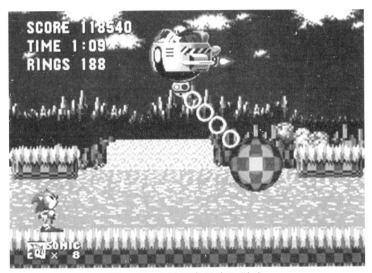

图 1-10 《刺猬索尼克》游戏

1.6 本章小结

本书的目的是使读者熟悉如何在自己工作的各种环境中应用强化学习技术。读者应熟悉诸如 TensorFlow 和 Keras 的深度学习框架,我们将利用这些框架部署许多深度学习模型来协同工作。虽然我们将花费一些时间来解释强化学习理论,并可能解释一些与深度学习重叠的知识,但本书的大部分内容将致力于讨论强化学习的理论和应用。因此,接下来我们将开始深入讨论强化学习的基础知识。

第 2 章
强化学习算法

读者请注意，本书将使用各种深度学习算法和强化学习算法。但是，由于我们的重点将转移到讨论这些算法的实现以及算法如何应用于产品环境中，因此必须花一些时间更详细地介绍算法的原理。本章的重点是引导读者逐步了解几种常用的强化学习算法示例，并以不同的问题为背景，利用 OpenAI Gym 环境来展示它们。

2.1　OpenAI Gym

在深入探讨任何具体示例之前，首先简要介绍一下本书大部分章节中将使用的软件。OpenAI 是位于旧金山湾区的一家研究机构。其在人工智能领域发表了大量论文，做出的最大开源贡献之一是 OpenAI Gym。OpenAI Gym 是针对 Python 语言发布的软件包，它提供了几种环境，用户可在该环境中使用各类强化学习算法。我们将专门针对电子游戏环境(在该环境中训练我们的算法)应用此软件包。下面，让我们首先尝试了解该软件包以及它的使用方法。

Gym 的基础是环境。在第 1 章中，我们已经讨论了环境、定义的各种变量以及环境的输出。在我们研究的每款游戏或环境中，通常会发现它们是不同的。本章中即将应对的车杆游戏是一味开胃菜。但是，稍后研究的《超级马里奥兄弟》游戏的环境将非常复杂。本章从了解车杆游戏和新环境开始，并尝试了解在该情景中需要做些什么来解决问题。车杆问题由 Barto、Sutton 和 Anderson(1983)在"Neuronlike Adaptive Elements That Can Solve Difficult Learning Control Problem"一文中描述。车杆问题的目的是使小车上的车杆保持平衡。对于车杆保持垂直状态的每一帧，将获得奖励值 1；但是，在任何给定帧，如果车杆不再保持垂直，则游戏将失败。下面我们将专注于如何应用策略梯度方法(强化学习的基石之一)，而不是专注用于求解此问题的方法。

2.2 基于策略的学习

基于策略的梯度方法侧重于直接优化策略函数，而不是尝试学习一个值函数(该值函数可产生关于给定状态下期望奖励值的信息)。简而言之，我们将在不利用值函数的情况下选择一个动作。策略分为以下两类：

- 确定性——一种将给定状态映射到一个或多个动作的策略，尤其是所采取的动作将带来"确定"结果的情况下。例如，在键盘上打字。当按 y 键时，可以确定字符 y 将出现在屏幕上。
- 随机性——一种能够在一组动作上产生概率分布的策略，从而存在一定概率，所采用的动作可能不是发生的动作。此策略专门用于环境不确定情况，并且是一种部分可观察马尔可夫决策过程(Partially Observable Markov Decision Process，POMDP)。

与基于值的方法相比，基于策略的方法具有一些特定优势，对于正在开展建模的读者而言，记住这一点非常重要。

首先，与基于值的方法相比，基于策略的方法倾向于收敛到更好的解决方案。其背后的原因是，我们由梯度引导至一个解决方案。直观地讲，梯度法指向微分函数的最陡方向。当运用误差函数并采用梯度下降法时，我们所采取的动作会使误差函数的值最小化(局部或全局)。因此，通常将得到一个可行的解决方案。相比之下，当动作间的差异较小时，基于值的方法产生的值相对差异更大或更敏感。也就是说，该方法无法保证收敛性。

其次，策略梯度法尤其擅长学习随机过程，而基于值的函数则不然。尽管并非每种环境都是随机的，但许多期望应用强化学习的实际示例是随机的。值函数之所以失败的原因在于，它们需要明确定义的环境，在该环境中它们内部的动作将产生确定性的特定结果。而随机环境不要求在采取相同动作时产生相同的结果，从而使得基于值的学习在这种环境中无效。相反，基于策略的方法不需要通过采取相同的动作来探索环境。具体来说，不需要平衡探索和利用("在结果已知情况下进行选择"与"尝试一种结果未知的动作")。

最后，基于策略的方法在高维空间中更显著有效，因为它们的计算成本明显降低，而基于值的方法要求我们为每一种可能的动作计算一个价值。如果空间中有大量动作(甚至无限个)，则几乎不可能收敛于一个解决方案，而基于策略的方法只是让我们执行动作并调整梯度。现在，读者已经对基于策略的学习方法有了基本了解，我们会将其应用到车杆问题。

2.3 策略梯度的数学解释

在对基于策略的方法有了大概了解之后，我们首先深入学习策略梯度的数学解释。读者应该记得，第 1 章简要介绍了马尔可夫决

策过程(MDP)的概念。我们将 MDP 定义为元组 (S, P, R, γ)，使得

$$R_s = \mathbb{E}[R_{t+1} | S_t = s],$$
$$G_t = R_{t+1} + \gamma R_{t+2} + \gamma^2 R_{t+3} + \cdots + \gamma^{k-1} R_{t+k}$$

定义了奖励和值函数后，现在可以讨论该策略的数学原理。agent 无法控制环境本身；但是，agent 肯定可以在一定范围内控制它执行的动作。这样，该策略被定义为在给定环境状态下所有动作的概率分布。其数学描述如下：

$$\pi(A_t = a | S_t = s),$$
$$\forall A_t \in A(s), S_t \in S$$

其中 π 表示策略，S 表示状态空间，A 表示动作空间，A_t 表示时间步 t 处的动作，S_t 表示时间步 t 处的状态。

现在，我们需要明白策略可以指导 agent 通过环境，即在环境的给定状态下执行特定的操作。在什么情况下适合使用策略梯度方法？策略梯度方法的目的是在 agent 拥有策略的情况下最大化预期奖励。因此，策略由 θ 参数表示，轨迹定义为 τ。轨迹含义为：在给定回合(episode)中，遵循给定策略时所观察到的动作、奖励和状态的序列。回合本身是指一种情景，在该情景中，agent 在到达某特定点(在该点处，我们达成问题目标或彻底失败)之前，持续在环境中执行某些动作[译者注，回合可以简单地理解为：agent 从一个特定状态出发，直到任务结束(达成问题目标或彻底失败)，被称为一个完整的回合]。因此，总奖励在数学上定义为 $r(\tau)$，使得

$$\arg\max J(\theta) = \mathbb{E}_\pi[r(\tau)]$$

然后，我们应用标准的机器学习方法，通过梯度下降法获得使得策略梯度最大化的参数。简要回顾一下，函数的梯度表示函数具有最大增长率的点，其大小是函数图形在该方向上的斜率。通常将

梯度乘以一个学习率，该学习率确定了收敛到函数最优解的速度。简而言之，梯度通常定义为给定函数的一阶导数。但是，如何利用它来优化我们选择的策略呢？

2.4 基于梯度上升的策略优化

基于梯度下降的优化方法常用于多种不同的机器学习算法中，例如线性回归以及用于多层感知器权重优化的反向传播。但是，这里将利用梯度上升来优化我们选择的策略。与尽可能最小化偏差不同的是，我们尽可能最大化通过整个回合(我们的算法应用于该回合)过程中获得的分数。因此，参数的更新如下所示。

$$\theta := \theta + \alpha \nabla_\theta J(\theta)$$

因此，问题的目标可以陈述如下。

$$\theta^* = \arg\max_\theta E_{\pi\theta}\left[\sum_t R(s_t, a_t)\right]$$

我们要尽可能选择这样的参数值，它们使得在给定状态下所采取动作获得的奖励最大化。对于正在建模的实例，我们尽可能选择能够最大化得分的神经网络权重。因此，预期总奖励的导数在数学上可定义为：

$$\nabla \mathbb{E}_\pi \lceil r(\tau) \rceil = \mathbb{E}_\pi \lceil r(\tau) \nabla \log \pi(\tau) \rceil,$$
$$\pi(\tau) = P(s_0) \prod_{t=1}^{T} \pi_\theta(a_t | s_t) p(s_{t+1}, r_{t+1} | s_t, a_t)$$

之所以采用几何和，是因为根据第 1 章马尔可夫决策过程中的定理，所采取的每个动作都是相互独立的。因此，应以类似的方式计算相关的累计奖励。在整个轨迹中重复此过程，该轨迹在逻辑上遵循给定回合的长度以及相关的奖励、状态和动作。对总奖励取对数，得到的式子如下。

$$\log \pi(\tau) = \log P(s_0) + \sum_{t=1}^{T} \log \pi_\theta (a_t | s_t) + \sum_{t=1}^{T} \log p(s_{t+1}, r_{t+1} | s_t, a_t),$$

$$\nabla \log \pi(\tau) = \sum_{t=1}^{T} \nabla \log \pi_\theta (a_t | s_t) \rightarrow \nabla \mathbb{E}_\pi [r(\tau)]$$

$$= \mathbb{E}_\pi \left[r(\tau) \left(\sum_{t=1}^{T} \nabla \log \pi_\theta (a_t | s_t) \right) \right]$$

分解该公式，预期奖励的对数可简化为每一单个奖励(单个奖励由策略针对给定时间、给定状态处的动作产生)的对数的和。了解这一点以及我们在强化学习中使用的通常被称为"无模型"算法的重要性在于，这些方程式隐式地表明了我们从未对环境建模的事实，因为我们从来不知道状态的分布。实际上，我们建模的唯一依据就是奖励。现在已解释了策略梯度的数学基础，接下来将其应用于一个经典控制问题：车杆。

2.5 使用普通策略梯度法求解车杆问题

为求解车杆问题，我们将利用 Keras 框架，这是一个以能够快速部署神经网络模型而闻名的库。尽管在本章后面将利用 TensorFlow 框架，但在此处部署的模型将是 applied_rl_python/neural_networks/models.py 文件中定义的软件包的一部分。在该文件中，用户将看到一些已经创建的类。无论是在本书内还是在本书之外的应用环境中，使用这些类将比从零开始编写代码容易得多。

```
class MLPModelKeras():
(Code redacted, please see the source code

    def create_policy_model(self, input_shape):
```

```
input_layer = layers.Input(shape=input_shape)
advantages = layers.Input(shape=[1])
hidden_layer = layers.Dense(n_units=self.n_units,
activation=self.hidden_activation)(input_layer)
output_layer = layers.Dense(n_units=self.n_columns,
activation=self.output_activation)(hidden_layer)

def log_likelihood_loss(actual_labels, predicted_
labels):
    log_likelihood = backend.log(actual_labels *
    (actual_labels - predicted_labels) + (1 - actual_
    labels) * (actual_labels - predicted_labels))
    return backend.mean(log_likelihood * advantages,
    keepdims=True)

policy_model = Model(inputs=[input_layer, advantages],
outputs=output_layer)

policy_model.compile(loss=log_likelihood_loss,
optimizer=Adam(self.learning_rate))
model_prediction = Model(input=[input_layer],
outputs=output_layer)
return policy_model, model_prediction
```

读者应该从这部分代码中读懂的事实是,该代码正在定义一个使用策略梯度方法的神经网络,着重指出的是,可以在以后出现的其他问题中重用或重新定义该神经网络。Keras 框架的优势在于,它使用户可以快速创建神经网络模型,如果读者使用过 TensorFlow,会发现 TensorFlow 的程序代码更加冗长,而 Keras 中额外的抽象层是自动化的,从而减少了编写神经网络模型所需的代码量。读者应查看图 2-1,以了解我们正在利用此神经网络解决的问题。

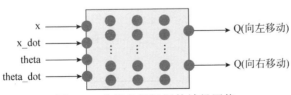

图 2-1 求解车杆问题的神经网络

输入层表示环境及其在给定状态下的方向,输出的两个类表示

21

我们可以采取的相应动作的概率。具体来说，我们将选择具有最高正确概率的动作，因为该问题被建模为一个分类问题。

接着，查看一下用于解决该问题的实际代码，该代码位于 chapter2/cart_pole_example.py 中。该文件的起始部分定义了一些有用参数，需要额外注意。Gym 的版本经常更新，本书使用的 Gym 版本是 0.10.5。在此版本中，我建议读者始终定义全局环境变量，然后在不同函数中访问环境的属性。此外，在此处定义 environment_dimension 变量会将环境重置为初始状态。现在，我们将注意力转移到 cart_pole_game() 函数，该函数完成本示例的大部分计算任务。具体来说，我们查看一下该代码的主体，该代码的主体包含了在某个特定回合中我们仍然没有输掉游戏前的所有代码。

```
state = np.reshape(observation, [1, environment_
dimension])
prediction = model_predictions.predict([state])[0]
action = np.random.choice(range(environment.action_
space.n), p=prediction)
states = np.vstack([states, state])
actions = np.vstack([actions, action])

observation, reward, done, info = environment.step(action)
reward_sum += reward
rewards = np.vstack([rewards, reward])
```

该代码的开头部分对于阅读过第 1 章中示例文件的读者来说应该很熟悉，但也存在一些细微差别。这里定义了一个 observation(观察)变量，每次实验开始时，该变量存储环境的初始状态。该模型产生的预测是概率性的，这里采取的特定动作是对我们可能采取的动作的随机抽样。然后将状态和动作附加到向量中，稍后将使用该向量。通常在给定环境中执行一个动作，将产生新的观察结果、当前的奖励以及一个指示器，指示器提示我们在该环境中是失败还是仍然继续。这个过程一直持续到输掉游戏为止，从而进入 calculated_discounted_reward() 函数。

2.6 什么是折扣奖励,为什么要使用它们

如前所述,策略梯度方法的目的是利用基于梯度最优化来选择一组动作,使得在给定目标情况下,这些动作在环境中实现最优结果。我们定义在给定状态下可以采取的动作的概率分布如下:

$$\pi_\theta(a|s) = P[a|s]$$

其中 π 为策略,θ 为参数,a 为动作,s 为状态。

由于这是基于梯度的优化问题,因此还需要定义代价函数,如下所示。

$$J(\theta) = E_{\pi\theta}[\Sigma\gamma r]$$

上面的公式是策略的得分函数(score function),它是我们选择的策略的预期/平均奖励。因为这是一种基于回合的任务,所以建议读者针对整个回合计算折扣奖励。下式给出了其计算方式:

$$J_1(\theta) = E_\pi \left[G_1 = R_1 + \gamma R_2 + \gamma^2 R_3 + \cdots + \gamma^{k-1} R_k \right]$$
$$J_1(\theta) = E_\pi \left(V(s_1) \right)$$

其中 k 为回合中的步数,G 为总折扣奖励,γ 为折扣调节参数,R 为奖励,V 为值。

calculate_discounted_reward()函数为产生的每个给定奖励声明一个折扣奖励向量,然后反向排列该向量,如下所示。

```
def calculate_discounted_reward(reward, gamma=gamma):
output = [reward[i] * gamma**i for i in range(0, len(reward))]
return output[::-1]
```

考虑在每个步骤中调节参数值被不同程度地提高,并且奖励是在该步骤我们采取动作的基础上由环境产生,因此我们对奖励进行

"打折"操作。然后对折扣奖励向量求取平均值,从而求得该回合代价函数的输出。

```
discounted_rewards -= discounted_rewards.mean()
discounted_rewards /= discounted_rewards.std()
discounted_rewards = discounted_rewards.squeeze()
```

读者将了解到我们对 discounted_rewards 向量执行的如下转换操作。该转换由 np.array.squeeze()函数完成,该函数将包含多个元素的数组作为输入参数,并对其中的元素执行连接操作,输出如下样式的结果。

$$[[1, 2], [2, 3]] \rightarrow [1, 2, 2, 3]$$

折扣奖励背后的原理非常简单,因为通过折扣奖励,使得原本为无穷的总和变得有限。如果不采取折扣奖励,这些奖励的总和将无限增长,因此将无法收敛于最优解决方案。

如何计算得分呢?

在我们的代码中,专门利用 score_model()函数计算,该函数使用训练过的模型运行由用户指定次数的实验,以得出这些次数实验的平均分数。这使得我们可以从总体上了解模型的运行情况,而不是查看一次由于偶然因素而使模型表现更好的实验。我们的得分函数也可以定义如下:

$$J(\theta) = E_\pi[R(\tau)]$$

其中 $R(\tau)$ 为预期的未来奖励。

该过程的实现方式非常简单。接下来我们解释一下 score_model()函数,如下所示。

```
def score_model(model, n_tests, render=render):
(代码经过节选,具体请参考 GitHub!)
            state = np.reshape(observation, [1, environment_
            dimension])
```

```
            predict = model.predict([state])[0]
            action = np.argmax(predict)
            observation, reward, done, _ = environment.
            step(action)
            reward_sum += reward
            if done:
                break
        scores.append(reward_sum)
    environment.close()
    return np.mean(scores)
```

读者将观察到,我们没有在每次计算模型得分时都进行环境的显示操作。我向读者推荐此方法,否则会显著降低训练过程速度,而且提供的是相对无用的信息。如果确实想要显示模型,则应该仅当认为达到了给定问题的基准时才这样做。

在该函数中,传递了一个模型作为参数,我们较早之前对该模型进行了一个批次(batch)的训练。通过利用各种状态及其相应折扣奖励,以及在每一种状态时我们采取的相应动作,对模型进行训练。直观地讲,我们尝试通过在每次迭代中选择一个随机动作来训练模型,使其更准确地预测出将导致特定奖励的动作。这样,权重将被优化,对于给定状态,每次都将产生一致的奖励。随着训练的深入,应该产生这样一个模型,当给定特定状态时,该模型将能够理解为了获得给定奖励而需要具体做什么。因此,根据问题框架,我们最终将产生一个模型,该模型将生成得分阈值,因为优化权重的目的是正确分类状态,以便随着时间的推进实现最大化得分的目标。

与所有梯度下降/上升问题一样,必须区分目标函数,以便能够计算梯度,从而用于优化权重。因为我们要区分的是一个概率函数,所以建议使用对数[这就是为什么我们使用一个对数似然损失(log-likelihood loss)作为误差函数,该函数定义在 neural_networks/models.py 文件中]。似然函数与该函数的对数似然函数的关系如图 2-2 和图 2-3 所示。

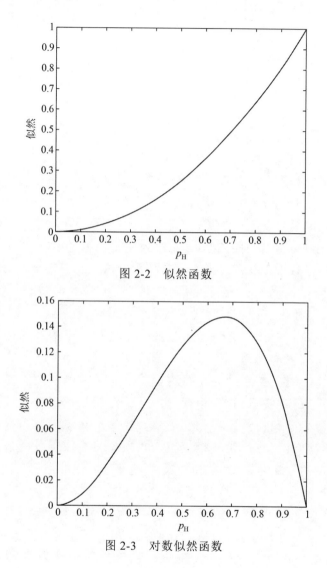

图 2-2 似然函数

图 2-3 对数似然函数

得分函数的导数由以下公式给出。

$$\nabla_\theta J(\theta) = E_\pi \left[\nabla_\theta (\log \pi(s, a, \theta)) R(\tau) \right]$$

由于利用了梯度上升策略,因此我们移动参数的方向最有可能最大化从环境中获得的奖励。

通过批训练更新参数之后,必须将状态、动作和奖励向量重新初始化为空。在详细讨论这一点之后,总结一下 cart_pole_game() 函数的工作原理,以下是处理流程。

(1) 初始化变量,这些变量通过在各自状态下与环境交互得到填充。

(2) 在给定回合中,执行动作直到游戏失败。在给定状态下,使用模型预测最佳动作来执行。将这些状态以及在这些状态下采取的动作和产生的奖励附加到向量中。

(3) 计算折扣奖励,然后使用这些奖励来训练一批次的状态、动作和奖励。

(4) 对训练好的模型进行评分,重复该过程直到收敛到用户指定的性能阈值为止。

充分解释以上代码之后,现在可以执行它并观察结果。当执行该代码时,应该看到如图 2-4 和图 2-5 所示的结果。

```
Episode: 50
Average Reward: 2.56
Score: 105.6
Error: 0.003579548

Episode: 100
Average Reward: 2.12
Score: 102.8
Error: 0.003846483

Episode: 150
Average Reward: 1.96
Score: 132.2
Error: 0.0039551817
```

图 2-4 策略梯度问题的示例输出

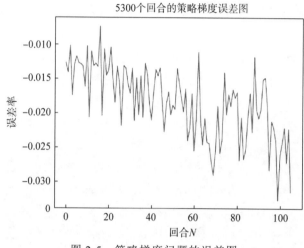

图 2-5　策略梯度问题的误差图

通过多次实验表明，当我们的目标设为 190 时，得到的解决方案在 5000~6000 个回合中收敛。现在，已经完成了一个回合问题(也是离散问题空间中的问题)的示例。我们现在知道了可以使用普通策略梯度(vanilla policy gradients)法求解的问题类型，那么，在什么情况下无法利用策略梯度法呢？

2.7　策略梯度的不足

在本阶段值得着重指出的对强化学习的较大批评之一是，策略梯度中存在的抽样效率问题。抽样效率是指，通过仅使用能够学习产生最重要信息的状态(译者注：抽取并使用重要的状态)，我们的算法能够达到更快学习的程度。具体来说，策略梯度不会区分一个回合中采取的各个动作。这意味着，如果在一个回合中采取的动作带来了很高的奖励，即使这些动作中的某些子集不是很理想，我们也可以得出结论，这些动作都是不错的。我们通常只能通过遍历非最佳策略来学习如何选择最佳策略。通过利用重要的抽样技术,已经减轻了其工作量;但是,这是离线(off-policy)学习中使用的一种技术,我们将在后面讨论。而且,此缺陷并非策略梯度法所独有。除此之外,策略梯度趋向于收敛到局部最大值而不是全局最大值。这也增加了训练出合格模型的难度。为了解决其中的一些问题,我们可以选择比前面所示的普通策略梯度中采用的回合模式的粒度更细的模式进行更新。这引导我们进入下一个主题,近端策略优化(Proximal Policy Optimization,PPO)。

2.8 近端策略优化(PPO)和 Actor-Critic 模型

PPO 通过对目标函数施加惩罚,然后在新改进的梯度下降上应用梯度策略,来专门处理策略梯度,从而免于陷于局部最大值。其公式如下所示:

$$\max_{\theta} \widehat{\mathbb{E}}_t \left[\frac{\pi_\theta(a_t|s_t)}{\pi_{\theta_{old}}(a_t|s_t)} \hat{A}_t \right] - \beta \widehat{\mathbb{E}}_t \left[\mathrm{KL} \left[\pi_{\theta_{old}}(\cdot|s_t), \pi_\theta(\cdot|s_t) \right] \right]$$

其中 β 为调节参数,KL 为 KL 散度,$\hat{A}_t$ 为优势函数。

这种适应性惩罚背后的基本原理是,我们利用了新旧策略之间的 KL 散度,对于一个回合内的每个迭代,该散度有所不同。如果 KL 散度值高于目标值 δ,则缩小调节参数。但是,如果它低于目标值 δ,则扩展搜索不同参数的区域。添加惩罚的好处在于,确保了用来搜索参数(搜索参数的目的是定义策略)的区域显著较小,并在比回合更小的粒度级别上基于正确度进行调整。这样,回合中的坏动作将直接受到惩罚,而不是被平均到其他决策中,这些决策有可能是不错的决策。这种按步骤而不是按回合的调整是 Actor-Critic 模型的核心思路,PPO 正是基于该模型。此时,与 KL 散度相关的调节参数是批评者模型,其中策略是参与者。

优势函数是 Actor-Critic 模型的关键组成部分，我们利用其代替算法决策过程中使用的值函数。这样做的理由是值函数具有更大的不确定性，而优势函数是更明显的凸函数。梯度优化背后的原理是，参数将朝着优势函数大于 0 的方向进行优化，并远离梯度小于 0 的方向。优势函数定义如下，它将用于取代值函数。

$$A(s, a)=Q(s, a) - V(s)$$

其中 $Q(s,a)$ 为状态 s 处执行动作 a 时产生的 Q 值，$V(s)$ 为状态 s 的平均值。

Actor-Critic 模型分为两种策略：Actor Advantage Critic(A2C)和 Asynchronous Advantage Actor-Critic(A3C)。之前简要描述过 Actor-Critic 模型，这两种算法的工作原理与其相同。但是，唯一的区别是，A3C 不同时(在每次迭代结束时)更新每个 actor 的全局参数，因此它是异步的。因此，A2C 的训练速度更快。

接着让我们更深入地了解该算法，将其应用到比车杆游戏更难处理的《超级马里奥兄弟》游戏中，并更直接地求解该问题。

2.9 实现 PPO 并求解《超级马里奥兄弟》

为实现该模型，本书将利用我们创建的某些软件包以及开源库中提供的代码。尽管可以更改游戏，但读者也应该尝试利用其他方式来解决该问题。由于 A3C 的训练时间稍长，本章将使用 A2C 算法。此外，本章还将向读者简要介绍如何设置 Google 云实例进行训练，建议将其用于诸如此类的任何基于强化学习的任务。

2.9.1 《超级马里奥兄弟》概述

《超级马里奥兄弟》(见图 2-6)是一款相对简单但经典的游戏，它使用户能够了解强化学习的独到之处，但同时又不增加复杂性(稍后将在本书其他游戏环境中看到这样的复杂性)。在 https://github.com/Kautenja/gymsuper-mario-bros/blob/master/gym_super_mario_bros/actions.py 中，列出了玩家可以应用的大量动作。

图 2-6 《超级马里奥兄弟》游戏截图

该游戏每一级关卡的目标都是相同的：尽量避开所有障碍物和敌人，以便能够在最后触摸旗杆，赢得关卡。旗杆将始终位于游戏每一级关卡的最右端，尽管玩家可以获得其他奖励，例如蘑菇和短暂无敌能力，但这些并不是主要目标。在该示例中，我们不考虑大多数用户的各自目标，因为这很可能导致模型难以训练，而得到的仅仅是次要奖励，我们的关键目标是到达旗杆。

2.9.2 安装环境软件包

对于此特定环境，鼓励读者使用 gym-super-mario-bros 软件包，可以使用以下命令进行安装。

```
pip3 install gym-super-mario-bros
```

《超级马里奥兄弟》不是 gym 包中提供的标准环境,因此需要创建一个环境。值得庆幸的是,gym-super-mario-bros 开源软件包可完成该任务,因此我们可以专注于该问题的模型架构。这次我们将直接使用 TensorFlow,而不是 Keras,但将访问 neural_networks/models.py 目录中的一个类。

2.9.3 资源库中的代码结构

与之前的示例不同,从现在开始,读者应该预料到将需要涉及模型的架构,因为模型定义在资源库的不同文件中(例如 neural_networks 目录和 algorithms 目录下)。在本示例中,代码的结构如下。

- A2C Actor-Critic 模型以一个类的形式定义在 models.py 中。
- algorithms/actor_critic_utilities 文件中包含 Model 类和 Runner 类。这些类,包括 ActorCriticModel 类,都在该文件中定义的 learn_policy()函数中实例化。该函数是大多数计算最终执行的地方。

这些类和函数来自 OpenAI 发布的基准库,并进行了简单修改。其背后隐藏的原理是,对于读者而言,了解这些模型的原理和工作方式非常重要,而不是简单调用它们,更不是从零开始编写代码。因此,我们首先讨论即将使用的模型及其原理。

2.9.4 模型架构

对于此问题,我们将其视为图像识别问题。因此,我们将使用一种简单的 LeNet 架构,它是一种卷积神经网络架构。卷积神经网络架构在图像识别领域应用非常广泛,最早由 Yann LeCun 在 20 世纪 80 年代后期开发。图 2-7 显示了一种典型的 LeNet 层次结构。

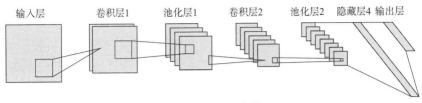

图 2-7 LeNet 架构

每一帧被视为一幅图片,对帧进行卷积操作以创建特征图,然后不断减少这些特征图的维数,直到获得 softmax 编码的输出向量,从向量中随机选取动作,然后最终对此批次进行训练,训练方式与前面的普通策略梯度示例中的类似。读者现在看到我们创建的 ActorCriticModel() 类的代码,该类包含了模型的架构和一些相关属性。

```
self.distribution_type = make_pdtype(action_space)
height, weight, channel = environment.shape
environment_shape = (height, weight, channel)
inputs_ = tf.placeholder(tf.float32, [None,
environment_shape], name="input")

self.distribution_type = make_pdtype(action_space)
height, weight, channel = environment.shape
environment_shape = (height, weight, channel)
inputs_ = tf.placeholder(tf.float32, [None,
environment_shape], name="input")
scaled_images = tf.cast(inputs_, tf.float32)/float(255)

layer1 = tf.layers.batch_normalization(convolution_
    layer(inputs=scaled_images,
filters=32,
kernel_size=8,
strides=4,
gain=np.sqrt(2)))
```

(代码稍后继续)

在介绍实现 Actor-Critic 模型并应用到《超级马里奥兄弟》游戏的关卡之前,简要讨论一下需要做哪些工作来预处理图像数据,以及数据在 CNN 中的移动方式。图像通常为 256 位,包含 3 个维度。

当将图像处理成 Python 矩阵时，意味着最初生成的矩阵的尺寸应为 $m \times n \times 3$，其中 m 和 n 分别为长度和宽度，矩阵的每一维代表一种颜色通道。具体来说，我们通常希望颜色通道分别代表红色、绿色和蓝色。在《超级马里奥兄弟》游戏示例中，我们希望该矩阵如图 2-8 所示。

```
array([[[104, 136, 252],
        [104, 136, 252],
        [104, 136, 252],
        ...,
        [104, 136, 252],
        [104, 136, 252],
        [104, 136, 252]],

       [[104, 136, 252],
        [104, 136, 252],
        [104, 136, 252],
        ...,
```

图 2-8 《超级马里奥兄弟》图像矩阵示例(预处理之前)

为了初步降低图像的复杂性，将对它们进行灰度处理，以使初始的三维矩阵变成一维矩阵。256 位中每一位分别表示颜色的亮度程度，其中 1 为黑色，256 为白色。由于 Python 数据结构的索引从 0 开始，255 是上限，因此将按比例缩放输入图像。现在我们专注于如何预处理数据，接着将进入第一个卷积层。

读者将注意到，在此处创建的层应用了一个函数，该函数使用一个包含在 TensorFlow 内嵌卷积层函数中的辅助函数。此外，在每个卷积层上使用了 batch_normalization()函数。如前所述，即将创建的特征图会不断缩小。理论上来讲，留下的数据是对分类目的最有用的像素。现在，继续向前，直到将所有特征图展平为一个数组，然后将其用于计算 $V(s)$。我们将该函数的输出以及其他一些重要的值定义为属性，在训练该模型期间将会调用它们。介绍完 ActorCriticModel 类，继续讨论 Model()类，其代码如下所示。

```python
class Model(object):
    def __init__(self, policy_model, observation_space, action_space, n_environments,
            n_steps, entropy_coefficient, value_coefficient, max_grad_norm):
```

(代码经过节选,具体请参考 GitHub!)

```python
        train_model = policy_model(session, observation_space, action_space, n_environments*n_steps, n_steps, reuse=True)
        error_rate = tf.nn.sparse_softmax_cross_entropy_with_logits(logits=train_model.logits, labels=actions_)
        mean_squared_error = tf.reduce_mean(advantages_ * error_rate)

        value_loss = tf.reduce_mean(mse(tf.squeeze(train_model.value_function) , rewards_))
        entropy = tf.reduce_mean(train_model.distribution_type.entropy())
        loss = mean_squared_error - entropy * entropy_coefficient + value_loss * value_coefficient
```

(代码稍后继续)

该代码从 policy_model() 开始,它实际上位于前面讨论的 ActorCriticModel() 类中。在实例化并执行完该类之后,我们从单次迭代中得到错误率(error rate),就像 Model() 类中所做的那样。下面将介绍如何使用 TensorFlow 进行标准的神经网络训练。接着看一下 Runner() 类。

```python
class Runner(AbstractEnvRunner):
    def __init__(self, environment, model, nsteps, total_timesteps, gamma, _lambda):
        super().__init__(environment=environment, model=model, n_steps=n_steps)

        self.gamma = gamma
        self._lambda = _lambda
        self.total_timesteps = total_timesteps

    def run(self):
        _observations, _actions, _rewards, _values, _dones = [],[],[],[],[]

        for _ in range(self.n_steps):
```

```
            actions, values = self.model.step(self.obs, self.
            dones)
            _observations.append(np.copy(self.observations))
            _actions.append(actions)
            _values.append(values)
            _dones.append(self.dones)
            self.observations[:], rewards, self.dones, _ =
            self.environment.step(actions)
            _rewards.append(rewards)
```

(代码稍后继续)

读者将观察到在之前(上一个示例)的代码片段中定义的一些变量。具体来说,定义了用作折扣因子的 gamma(伽玛)。再次提醒,对于梯度下降法来说,当梯度较小时,比起具有较大梯度值的神经网络,权重优化将容易得多。当利用允许在该环境中执行的最大步骤数遍历每一个迭代时,会对 observations、actions、values、rewards 和布尔值执行 append(附加)操作,该布尔值用于确定是否失败或仍在进行当前回合游戏。

```
(代码经过节选,具体请参考 GitHub)
delta = _rewards[t] + self.gamma * nextvalues *
nextnonterminal - _values[t]
            _advantages[t] = last_lambda = delta + self.gamma *
self._lambda * nextnonterminal * last_lambda

        _returns = _advantages + _values
        return map(swap_flatten_axes, (_observations, _actions,
_returns, _values))
```

在该代码中,我们转移到函数的末尾部分,在此处计算了 delta 值,即 rewards、lambda、returns 等值在各个步骤之间的差。最终转向 train_model()函数,如下所示。

```
    model = ActorCriticModel(policy=policy,
            obsevration_space=observation_space,
            action_space=action_space,
            n_environments=n_environments,
n_steps=n+steps,
entropy_coefficient=entropy_coefficient,
value_coefficient=value_coefficient,
```

```
max_grad_norm=max_grad_norm)
model.load("./models/260/model.ckpt")
runner = Runner(environment,
                model=model,
                n_steps=n_steps,
                n_timesteps=n_timesteps,
                gamma=gamma,
                _lambda=_lambda)
```

(代码经过节选,具体请参考 GitHub)

读者已经了解了这些函数,现在根据在文件头以及 train_model() 函数中定义的超参数对它们进行实例化。从现在开始,在训练模型方面,读者看到的过程应与之前示例中的过程相似。现在,已经对该示例进行了适当的概述。下面介绍训练这样一个模型所面临的挑战以及我们观察到的结果。

2.10 应对难度更大的强化学习挑战

强化学习领域的车杆问题和其他一些经典控制问题相对容易,因为无论选择哪种方法,都不会花费太多的时间就可得到最佳解决方案。但是,对于更抽象的问题,尤其是与本节示例类似的问题,训练时间可能会呈指数级增加。例如,有些已应用于《刺猬索尼克》游戏的 A2C 和 A3C 算法的实现,花费 10 小时仍无法完成一级关卡的训练。尽管在上面的《超级马里奥兄弟》游戏示例中没有呈现出一些复杂之处,但应牢记这一点。因此,对于这样的问题,需要使用云解决方案。虽然稍后将讨论 AWS 以及如何应用它,但我认为对于读者来说,学习其他框架也很重要。因此,我们将使用 Google 云。作为额外奖励,它们仍然为新用户提供免费积分,这将大大简化此代码的应用。

所有数据科学家或机器学习工程师都将认同这一点:他们需要的解决方案应该可应用云资源进行产品化和实验。AWS 和 Google

云是读者应熟悉的两类云解决方案，这不仅仅因为它们对于代码的产品化有意义。Google 云仪表盘示例如图 2-9 所示。

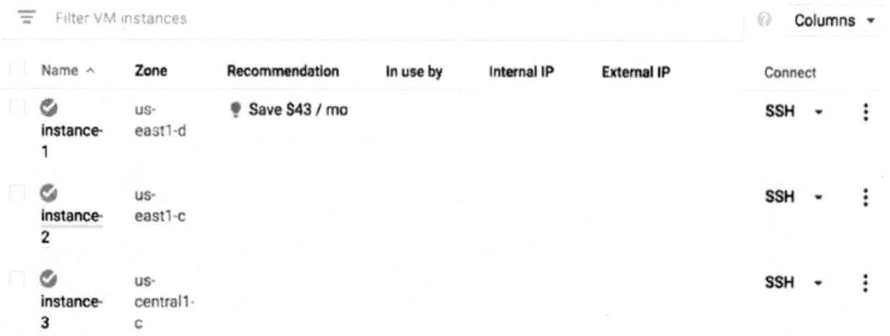

图 2-9　Google 云仪表盘示例

读者会期望在单击 SSH 图标时加载一个(这里应该是 Linux)终端，这将需要一些标准安装(安装 Git、不同的 Python 软件包等)。用户在终端所做的一切与他们在本地计算机上所做的没有不同。但是，如果用户正在使用 Linux 操作系统，则语法上会有一些差异。

注意：本节的重点是理解应该在云资源上而不是在本地计算机上训练诸如此类的解决方案。

现在看一下实际运行该游戏的主函数。

```
def play_super_mario(policy_model=ActorCriticModel,
environment=environment):
    (代码经过节选，具体请参考 GitHub！)
    observations = environment.reset()
    score, n_step, done = 0, 0, False

    while done == False:

        actions, values = model.step(observations)
        import pdb; pdb.set_trace()

        for action in actions:
```

```
        observations, rewards, done, info =
        environment.step(action)
        score += rewards
        environment.render()
n_step += 1

    print('Step: %s \nScore: %s '%(n_step, score))
    environment.close()
```

在最后一段代码中,使用了所有必要的类。我们在这里讨论的最后一部分是如何顺利实现训练过程。为此,建议用户熟悉一下 docker 容器知识。

2.11 容器化强化学习实验

在训练一个强化学习 agent 时,你可能不想坐下来紧盯着 agent 通过优化其策略来熟悉环境,并且你肯定会在训练它的大量时间中仍然需要使用计算机。这就是利用云资源的原因。但是,仅在云环境上运行应用程序是不够的。在 AWS 或 Google 云上,如果不在后台运行进程,则在连接断开时(由于计算机死机等原因)将失去所有进度,不得不从最后一个检查点处或从头开始,具体取决于你是否修改了代码以保存某些检查点。因此,利用 docker 容器技术很重要。

docker 容器是一种有意义的解决方案,利用它可创建从终端运行的应用程序虚拟环境。简而言之,可以创建一个虚拟的"实例",从该虚拟环境中能够快速启动应用程序并运行它。另一个额外好处是 docker 包含几条命令,这些命令可帮助用户运行诸如此示例的进程并在进程停止时重启它。在我们正在此处执行的任务背景下,一旦觉得 agent 的训练时间足够长,就可以终止该进程,然后检查 agent 的进度,在认为必要时返回训练。首先,看一个 Docker 示例文件。

图 2-10 是一个模拟的 Docker 文件,其中我们将看到三类命令。

具体来说，它们是 FROM、COPY 和 RUN。FROM 定义了我们要在其中运行该容器的 Python 版本。尽管本书中存在一些使用 Python 2 的示例，但所有示例都应与 Python 3 兼容，且在 2020 年以后将不再支持 Python 2。接着，COPY 命令指出了我们要使用的资源库中的特定文件。最后是 RUN 命令，这里安装所需的 Python 软件包。

```
FROM python:3

COPY .git /.git
COPY __init__.py .
COPY repo/ repo
COPY repo2/ repo
COPY repo3/ repo

RUN pip install pandas numpy
```

图 2-10　Docker 示例文件

注意：需要重点注意的是，在实例化新容器时，必须在 Docker 文件中指出所有必需的文件、资源库和 Python 模块。如果不这样做，docker 容器将无法执行代码。

通常使用以下命令创建容器。

```
"sudo docker -t build . [container name] . "
```

假设已安装 docker，且没有缺失需要复制的任何文件，则该命令将会按指定的名称创建一个 docker 容器。执行后，建议用户运行以下命令来启动文件。

```
"sudo docker run --dit --restart-unless-stopped python3 -m path.
  to.file"
```

2.12 实验结果

本示例(《超级马里奥兄弟》示例)主要是出于演示目的,但是,在求解难度更大的强化学习问题时,强调一点很有意义——agent 必须训练大量的时间。与一些较简单的机器学习示例不同,本示例的训练将花费很长时间才能有效,这点与基于深度学习的自然语言处理难题更相似。在本示例中,agent 常常会耗光时间,因为它卡在了障碍物上(例如相对较早出现的管道),或者运气不佳,被诸如蘑菇(goomba)的敌方战士相对迅速地杀死。当对 agent 进行了 5 小时的训练后,通常会观察到它的性能显著变好,最明显的事实是,在遇到空间中的任何敌人时可以避免死亡。但是,当它遇到障碍物且卡住时,就不太可能回退以找到替代的前进道路。最成功的 agent 是那些受过 12 小时以上训练的。但是,该解决方案通常还没有完成,也不一定是完美的。agent 的成功似乎大部分都取决于它在关键时刻所采取的动作,尤其是正确时机的跳起,并且它倾向于避免杀死敌人,而是偏爱尽量不掉入关卡的陷阱中。在某些情况下,马里奥可以获胜;但是,需要重点注意的是,这是较简单的游戏关卡之一。

2.13 本章小结

阅读完本章之后,读者应该对应用基于回合和时序差分方法的一些基本和较高级类型的强化学习算法感到得心应手。本章主要内容如下:
- 了解面临的问题类型——与大多数机器学习问题类似,对于不同类型的数据,可使用不同的模型。正在处理的是大型状态空间吗?任务是回合型的吗?如果不是,是否真正地希望/需要基于更精细步骤的学习算法?在得到解决方案之前,请花点时间思考这些问题。
- 对于一些难解之题,训练强化学习解决方案是非常耗时的,因此需要利用云资源进行训练。例如,对于类似一些高级自然语言处理问题,读者会发现本地机器不适合训练模型。尽管在本地计算机上编写大多数代码是很有意义的,但是请设法在其他地方使用这些代码。

随着第一种算法的完成,我们将继续致力于探讨不同的基于值的方法,例如 Q 学习和深度 Q 学习。在第 3 章,将再次采用相同的模式,先处理一个较简单问题,然后再在一个更大的环境中处理一个更复杂的问题。

第 3 章
强化学习算法：Q 学习及其变种

在完成策略梯度和 Actor-Critic 模型的初步探讨之后,现在继续讨论对读者有用的其他深度学习算法。具体来说,本章将讨论 Q 学习、深度 Q 学习以及深度确定策略梯度。学习完这些内容后,读者掌握的知识可支撑处理更抽象、针对特定领域的问题,这些问题将教会读者如何将强化学习技术应用于不同的任务。

3.1　Q 学习

Q 学习从属于无模型(model-free)学习算法族。该算法通过遍历所有可能的动作并评估每个动作来学习策略。在该算法中,会频繁访问两个矩阵：Q 矩阵和 R 矩阵。前者与该算法名同名,包含了用于实现策略的环境累积知识。Q 矩阵中的所有项都被初始化为 0,

目标是使所产生的奖励最大化。在环境的每个步骤中，都会更新 Q 矩阵。R 矩阵的每一行代表一个状态，而各列代表行状态向另一状态迁移的奖励。该矩阵的结构类似于相关矩阵(correlation matrix)，其中行索引和列索引互成镜像。图 3-1 和图 3-2 分别给出了 Q 矩阵和 R 矩阵的可视化示例。

$$Q = \begin{array}{c} \\ 0 \\ 1 \\ 2 \\ 3 \\ 4 \\ 5 \end{array} \begin{array}{c} 0 \quad 1 \quad 2 \quad 3 \quad 4 \quad 5 \end{array} \\ \left[\begin{array}{cccccc} 0 & 0 & 0 & 0 & 0 & 0 \\ 0 & 0 & 0 & 0 & 0 & 100 \\ 0 & 0 & 0 & 0 & 0 & 0 \\ 0 & 80 & 0 & 0 & 0 & 0 \\ 0 & 0 & 0 & 0 & 0 & 0 \\ 0 & 0 & 0 & 0 & 0 & 0 \end{array} \right]$$

图 3-1　Q 表的可视化示例

$$R = \begin{array}{c} 状态 \\ 0 \\ 1 \\ 2 \\ 3 \\ 4 \\ 5 \end{array} \begin{array}{c} 动作 \\ 0 \quad 1 \quad 2 \quad 3 \quad 4 \quad 5 \end{array} \\ \left[\begin{array}{cccccc} -1 & -1 & -1 & -1 & 0 & -1 \\ -1 & -1 & -1 & 0 & -1 & 100 \\ -1 & -1 & -1 & 0 & -1 & -1 \\ -1 & 0 & 0 & -1 & 0 & -1 \\ 0 & -1 & -1 & 0 & -1 & 100 \\ -1 & 0 & -1 & -1 & 0 & 100 \end{array} \right]$$

图 3-2　R 表的可视化示例

agent 可以通过查看 R 表(R table)，找到其可以采取的立即动作，但是除此之外它看不到其他任何内容。正因为有此限制，这也是 Q 表(Q table)存在的必要性所在。如前所述，Q 表包含了给定时间段内填充的关于环境的所有累积信息。从某种意义上讲，可以将 Q 表视为地图，将 R 表视为世界。具体来说，Q 表的更新方式如下。

$$Q(s_t, a_t) \coloneqq Q(s_t, a_t) + \alpha \left[\left(r(s_t, a_t) + \gamma \cdot \max \{ Q(s_{t+1}, a_{t+1}) \} \right) - Q(s_t, a_t) \right]$$

其中 $Q(s_t, a_t)$ 为矩阵的单元项，α 为学习率，γ 为折扣因子，$\max \{Q(s_{t+1}, a_{t+1})\}$ 为最大的 Q 表值。

3.2 时序差分(TD)学习

在第 1 章中,简要谈到了马尔可夫决策过程的主题。更具体地说,MDP 指的是部分随机的事件,但也依赖于决策者的控制。我们将 MDP 定义为如下 4 元组:

$$(S, A, P_a, R_a)$$

其中 S 为状态集合,A 为可采取的动作集合,P_a 为时间 t 处的状态 s 执行动作 a 后迁移到时间 $t+1$ 处的状态 s' 的概率,R_a 为因动作 a 导致从状态 s 迁移到状态 s' 后获得的立即奖励。

作为示例,图 3-3 是一个马尔可夫决策过程的示意图。

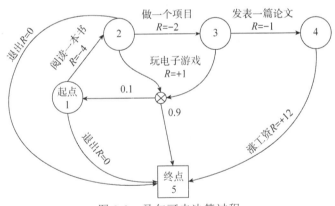

图 3-3 马尔可夫决策过程

如前所述,大多数强化学习都是围绕状态进行的,可以根据这些状态执行产生奖励的动作。我们的终极目标是为决策者选择能够最大化奖励值的最优策略。第 1 章中简要提到了时序差分学习(TD),现在是时候详细讨论该主题了。

TD 学习通常被描述为预测出一个数值(quantity)的方法,该数值取决于特定信号的未来值。它指的是在不同时间步上预测的"时序差分"。TD 学习的设计思路为，更新当前时间步的预测，以便紧接着下一时间步的预测更正确。Q 学习本身是 TD 学习的一个特例。尤其需要在这里介绍的是 epsilon-greedy 算法，该算法是用于求解 TD 学习问题的一种方法。

3.3 epsilon-greedy 算法

经过大量迭代，Q 表的质量最终足够好，可被 agent 直接使用。为了达到此目的，我们希望 Q 学习算法利用的表中信息少于它探索的信息。总的来说，这就是探索与利用之间的折中，它由 epsilon 参数控制。此处的关键点是，可能被采用并达成解决方案的第一条路径不一定是最好路径。根据该陈述(反之亦然)，如果我们继续搜索，未必总是会找到比当前解决方案更好的解决方案，因此我们放弃继续求解该问题。为了缓解这一难题，建议使用 epsilon-greedy 算法。

epsilon-greedy 算法从属于多臂老虎机(multi-armed bandit)问题族范畴。该问题描述的是，通过在各种选项中进行选择，最终目标是获得最大奖励。此问题的经典示例是，假设某个赌场中有四台机器(老虎机)，每台机器具有不同的未知奖励概率。伯努利(Bernoulli)多臂老虎机可描述为一组动作(action)和奖励(reward)，以元组形式表示为<A, R>，其中存在 k 台机器，它们的奖励概率为$\{\theta_1, ..., \theta_k\}$。每个动作对应于与相应老虎机的一个交互，并且奖励是随机的，返回值是概率值 $Q(a_t)$ 或 0。预期奖励表示为以下公式：

$$Q(a_k) = \mathbb{E}[r_k | a_k] = \theta_k, k \in \{1, ..., k\}$$

我们的目标是通过选择一系列最优动作来最大化累积奖励，其中最优奖励概率和损失函数分别由以下公式给出。

第3章 强化学习算法：Q学习及其变种

$$\theta^* = Q(a^*) = \max_{a \in A} Q(a) = \max_{1 \leq i \leq k} \theta_i$$

$$L_T = \mathbb{E}\left[\sum_{t=1}^{T}(\theta^* - Q(a_t))\right]$$

尽管存在多种方法可以求解多臂老虎机问题，但这里将聚焦于利用该策略。这是一种通过以下公式估算动作质量的算法。

$$\hat{Q}_t(a) = \frac{1}{N_t(a)} \sum_{\tau=1}^{t} r_\tau \, \mathbb{1}\,[a_\tau = a]$$

其中 $N_t(a)$ 为动作 a 被选取的次数，$\mathbb{1}$ 为二元指示函数。

如果值较小，则将探索紧邻环境(immediate environment)。否则，将利用目前已知的最好可能动作。为了演示整个 Q 学习算法，我们将学习如何玩一个名为《冰湖》的电子游戏。

3.4 利用 Q 学习求解冰湖问题

《冰湖》是 Gym 中提供的一款游戏，其中玩家尝试训练 agent 从湖上的起点移动到终点，从而穿越湖面。但是，并非所有冰块都是冻结的，如果踩到非冻结冰块，则会输掉比赛。除非到达目的地，否则不会获得任何奖励。读者可以想象图 3-4 所示的环境。

S	F	F	F
F	H	F	H
F	F	F	H
H	F	F	G

图 3-4 《冰湖》游戏环境

与我们编写的大多数其他代码类似，首先定义了后面可以使用的参数以及环境。两个主要函数 populate_q_matrix() 和 play_frozen_lake() 中包含了一些之前定义的辅助函数。下面首先看一下用于填充 Q 矩阵的函数。

```
def populate_q_table(render=False, n_episodes=n_episodes):
(文档经过节选,具体请参考 GitHub!))
    for episode in range(n_episodes):
        prior_state = environment.reset()
        _ = 0
        while _ <max_steps:
            if render == True: environment.render()
            action = exploit_explore(prior_state)
            observation, reward, done, info = environment.
                                                    step(action)

            update_q_matrix(prior_state=prior_state,
                            observation=observation,
                            reward=reward,
                            action=action)
```
(代码稍后继续)

浏览代码到第二个辅助函数 update_q_matrix() 处,会看到定义了多个 episode(回合),我们将在这些回合中填充 Q 表。读者可以或多或少地添加回合,以了解程序性能如何变化,这里我们选择了 10 000 个回合。现在查看第一个辅助函数 exploit_explore()。该函数的算法过程简单明了,它执行 epsilon-greedy 探索算法,来确定我们应该采取的两个动作。该函数代码如下所示。

```
def exploit_explore(prior_state, epsilon=epsilon):
(代码经过节选,具体请参考 GitHub)
    if np.random.uniform(0, 1) < epsilon:
        return environment.action_space.sample()
    else:
        return np.argmax(Q_matrix[prior_state, :])
```

正如读者所看到的,如果从均匀分布中随机抽取的值为 0,则仅探索一个随机动作。否则,选择在给定状态下已知的最优动作。向后继续查看更大的函数的代码,像前面示例中所做的那样,使 agent 在环境中执行动作。接着情况有所不同,我们必须更新 Q 矩阵。

```
def update_q_matrix(prior_state, observation , reward, action):
prediction = Q_matrix[prior_state, action]
    actual_label = reward + gamma * np.max(Q_
    matrix[observation, :])
    Q_matrix[prior_state, action] = Q_matrix[prior_state,
    action] + learning_rate*(actual_label - prediction)
```

根据之前的公式更新 Q 矩阵的项,其中矩阵的每一列代表一个要采取的动作,每一行代表一个不同的状态。在每个回合中会继续执行该过程,直到达到允许执行的最大步骤数或跌下冰面。一旦达到最大回合数,就可以使用获得的 Q 表来玩游戏了。读者在终端运行游戏时,应该能够观察到如图 3-5 所示的结果。

```
Episode: 0

SFFF
FHFH
FFFH
HFFG

Episode: 1
  (Left)
SFFF
FHFH
FFFH
HFFG

Episode: 1
  (Left)
SFFF
FHFH
FFFH
HFFG
```

图 3-5 《冰湖》游戏

当在给定回合中获胜或失败时,终端将输出一些信息。我们通常会使用提供的参数来多次实验并观察,agent 通常会在 10 回合中获胜 2 到 3 次，并且将在 20～30 步内得到解决方案。

在某种程度上 Q 学习的主要优势是它不需要模型，并且该算法相当透明。很容易解释清楚为什么处于给定状态的 agent 会选择某个动作。尽管如此，它的主要缺点是，如果我们想要利用信息充分填充 Q 矩阵，则在处理非常复杂的环境时，在给定状态下，获得关于执行什么动作的知识所需要的经验在计算上开销巨大。然而，《冰湖》游戏示例有相当大的局限性，对于更复杂的电子游戏环境，很可能会花费非常长的时间才能获得较好的 Q 表。为了克服此局限，设计了深度 Q 学习。

3.5 深度 Q 学习

深度 Q 学习(DQL)很大程度上源于 Q 学习，这两种方法之间的唯一真正区别是，深度 Q 学习近似处理其 Q 表中的值，而不是尝试手动填充它们。如何精准地完成该任务，需要综合运用 epsilon-greedy 搜索(或另外一种替代算法)与动作执行后的结果。epsilon-greedy 搜索算法帮助我们判断是利用 Q 表还是继续探索，然后基于该状态下动作的输出值更新 Q 矩阵。从这个意义上讲，可以看到，我们希望将"达到我们的目标"和"采取的动作"之间的损失(loss)降到最低。因此，现在具有利用梯度下降的基础，它由以下公式表示。

$$\mathfrak{L}_i(\theta_i) = \mathbb{E}_{a\sim\mu}\left[\left(y_i - Q(s,a;\theta_i)\right)^2\right],$$

$$y_i := \mathbb{E}_{a'\sim\mu}\left[r + \gamma \max_{a'} Q(s',a';\theta_{i-1}) \mid S_t = s, A_t = a\right]$$

其中，μ 为行为策略，θ 为神经网络参数。

目标标签和 Q 矩阵分别由两个独立的神经网络预测。目标网络共享 Q 网络的权重和偏差，但是它们在 Q 网络之后进行更新。接着，我们讨论经验回放(experience replay)的重要性以及在这里如何利用它。如果在强化学习应用背景中引入全新数据，则需要重构神经网络的权重。因此，这就是为什么经常要针对不同应用目的训练不同模型的原因。经验回放是我们通过保存观察到的经验以便利用它们，从而有助于减少可能观察到的经验之间的相关性。实际上，我们在内存中存储了本章起始部分介绍的元组。在训练过程中，使用该元组计算目标标签，然后应用梯度下降法，获得作用于整个环境的权重和偏差。接着，继续前行，现在尝试使用深度 Q 学习解决问题，看看问题的复杂性如何显著变化。

3.6 利用深度 Q 学习玩《毁灭战士》游戏

应用深度 Q 学习的经典示例之一是早期版本的《毁灭战士》游戏，如图 3-6 所示，它的环境也是测试各种机器学习算法的绝佳环境。《毁灭战士》是第一人称射击游戏，玩家在三维环境中四处游走，并与敌方战士作战。由于这是一款较老的 3D 游戏，因此玩家可以在环境中四处移动，就像我们许多理论上的 agent 在 Q 矩阵中所做的一样。这将是我们应用强化学习的第一种连续控制问题。

简而言之，区分连续控制系统和离散控制系统的方式是，前者中的变量和参数是连续的，而后者中的是离散的。强化学习应用背景中的连续过程示例包括驾驶汽车或教机器人走路。离散控制过程示例包括我们已经解决的第一个问题——车杆游戏，以及"经典控制"问题中的一些其他问题，如钟摆问题。为了理解算法，尽管存在许多值得分析的离散任务，但是许多可利用强化学习实现的任务是连续的。该游戏存在庞大的状态空间，这使得深度 Q 学习成为理想选择。对于该问题，我们将查看简单级别与较难级别的差异，并查看算法性能方面的差异。

图 3-6 《毁灭战士》游戏的某个级别关卡示例

具体到游戏本身，目标相当简单明了。我们必须在不死的情况下完成关卡，而这显然需要杀死通向关卡终点的所有敌方战斗人员。大多数敌人会首先发招，因此算法将主要集中在基于此的反应方式训练上。总体而言，基于该算法执行的两个主要过程是：(1)对环境进行采样并将经验存储在 MDP 元组中；(2)选择其中一些经验用于批训练示例。除了讨论使用哪种类型的模型架构外，让我们首先讨论如何预处理模型的数据。

```
class DeepQNetwork():

    def __init__(self, n_units, n_classes, n_filters, stride,
    kernel, state_size, action_size, learning_rate):
     (代码经过节选，具体请参考 GitHub!)

     self.input_matrix = tf.placeholder(tf.float32, [None,
     *state_size])
     self.actions = tf.placeholder(tf.float32, [None])
     self.target_Q = tf.placeholder(tf.float32, [None,
     *state_size])
```

第3章 强化学习算法：Q学习及其变种

```
self.network1 = convolution_layer(inputs=self.input_matrix,
                                  filters=self.n_filters,
                                  kernel_size=self.kernel,
                                  strides=self.stride,
                                  activation='elu')
```
(代码经过节选，具体请参考 GitHub!)

与之前的 TensorFlow 图类似(我们已经将其定义为 graphs)，将从定义几个特定属性开始。这些属性稍后将在 doom_example.py 中的 play_doom()函数中使用，稍后我们将讨论该问题。继续分析代码，可以看到，与《超级马里奥兄弟》游戏示例类似，这里将使用 LeNet 架构，但在此示例中，由于要处理图像帧，因此将利用一个可接受四个维度的层。同样，最终将特征图展平为一个数组，然后通过一个全连接的 softmax 层输出它。从该 softmax 层开始，将在训练过程中采样动作。图 3-7 显示了我们将用于深度 Q 网络的模型架构示例。

图 3-7 深度 Q 网络架构示例

继续讨论输入数据，在之前的示例中，我们没有对帧进行堆叠处理，而是将当前状态和先前状态的数据进行格式重排，并传递重排后的矩阵。数据预处理和堆叠帧之所以重要，特别是在三维环境中，是因为它使得深度 Q 网络能够了解 agent 的诱导动机。此方法由 Deep Mind 提出。我们通过以下函数对帧进行预处理和堆叠：

```
def preprocess_frame(frame):
    cropped_frame = frame[30:-10,30:-30]
```

```
normalized_frame = cropped_frame/float(255)
preprocessed_frame = transform.resize(normalized_frame,
 [84,84])
return preprocessed_frame
```

首先利用一幅灰度(grayscaled)图像，感谢 vizdoom 库以该格式提供给我们。如果不是灰度图像，则应利用 OpenCV 之类的库来执行该预处理。继续分析，基于《超级马里奥兄弟》示例中的相同理由，再次利用 255 缩放像素值。但是，稍有差别的是，在本示例中，我们将剪裁帧的顶部，因为《毁灭战士》游戏的顶部仅用于显示天空，不包含任何有价值信息。这里利用前面的函数对帧执行堆叠操作：

```
def stack_frames(stacked_frames, state, new_episode, stack_size=4):
    frame = preprocess_frame(state)

    if new_episode == True:

        stacked_frames = deque([np.zeros((84,84), dtype=np.int)
        for i in range(stack_size)], maxlen=4)
        for i in range(4):
            stacked_frames.append(frame)

        stacked_state = np.stack(stacked_frames, axis=2)

    else:

        stacked_frames.append(frame)
        stacked_state = np.stack(stacked_frames, axis=2)

    return stacked_state, stacked_frames
```

与将帧转换为四个堆栈的函数不同，该函数的重点在于其工作的准确程度。首次调用该函数时，接受前四帧。继续执行，添加最新帧的同时删除最老帧，也就是一个先进先出(FILO)过程。但是要记住的是，该过程不代表现实情况，因为人类看到的不是多帧交错出现，而是一次看到所有帧。除此之外，由于需要使用内存存储这些堆叠图像，使得训练变得更加困难。在学习接下来章节中的不同

第 3 章 强化学习算法:Q 学习及其变种

示例时,读者应牢记这一点。继续分析,我们将使用稍微复杂一些的 epsilon-greedy 策略,其中还将使用衰减率(decay rate)参数,如以下函数所示。

```
def exploit_explore(session, model, explore_start, explore_
stop, decay_rate, decay_step, state, actions):
    exp_exp_tradeoff = np.random.rand()
    explore_probability = explore_stop + (explore_start -
    explore_stop) * np.exp(-decay_rate * decay_step)

    if (explore_probability > exp_exp_tradeoff):
        action = random.choice(possible_actions)
    else:
        Qs = session.run(model.output, feed_dict = {model.
        input_matrix: state.reshape((1, * state.shape))})
        choice = np.argmax(Qs)
        action = possible_actions[int(choice)]
```

该 epsilon-greedy 策略背后的原理与我们在原始的 Q 学习示例中看到的基本相同,不同之处在于其衰减是呈指数级的,因为随着时间的推移,探索到的信息更可能变得越来越少,从而迫使算法利用其积累的知识。在解释了辅助函数之后,现在逐步分析实际上用于训练模型的函数。闲话少说,我们查看一下在本级关卡上模型的训练结果。然后,将处理另一级别关卡,看看模型运行效果如何。

简单的《毁灭战士》关卡

在此应用背景中,玩家将处于一个简单环境中,他们可以向左移动、向右移动或向敌方战士开枪射击。而敌方战士不会还枪,只是简单地向左或向右移动。读者在运行代码时,输出内容和屏幕显示应该如图 3-8 和图 3-9 所示。

```
Episode: 3 Total reward: 94.0 Explore P: 0.9701
Episode: 4 Total reward: 27.0 Explore P: 0.9644
Episode: 13 Total reward: 76.0 Explore P: 0.8893
Episode: 14 Total reward: 91.0 Explore P: 0.8884
Episode: 17 Total reward: 95.0 Explore P: 0.8705
Episode: 20 Total reward: 27.0 Explore P: 0.8485
```

图 3-8　训练模式的屏幕截图

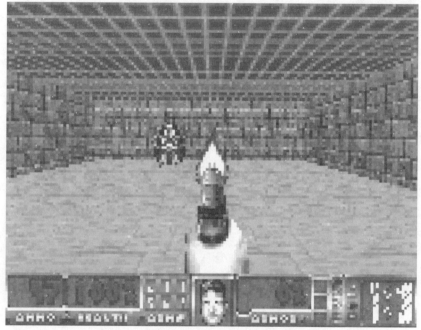

图 3-9　简单的《毁灭战士》环境示例

3.7　训练与性能

　　图 3-10 显示了各回合中 Q 矩阵的训练结果以及测试结果。

　　读者应该意识到，类似这样的任务，由于需要执行预处理和计算，它们将占用相当大的内存，这点前面我们也已经谈到过。除此之外，有时神经网络会陷入局部最优状态，从而无法正确学习要采取的动作。尽管列出的参数通常会产生可接受的测试解决方案，但有些情形这种神经网络的效果也一般。这是其局限性之一。

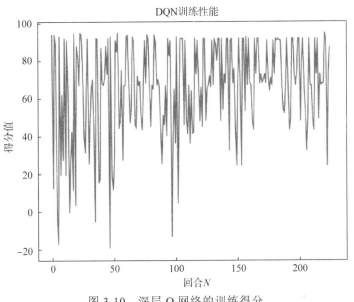

图 3-10　深层 Q 网络的训练得分

3.8　深度 Q 学习的局限性

正如前面所介绍的，深度 Q 学习并非没有缺点。但是，除了本示例外，这些不足中的大多数往往会出现在哪里呢？塞巴斯蒂安·特伦(Sebastian Thrun)和安东·施瓦茨(Anton Schwartz)在他们 1993 年的论文"Issues in Using Function Approximation for Reinforcement Learning"中对此进行了专门研究。他们发现，由于高估的存在，深度 Q 网络经常学到非常高的动作值。原因归结于以下目标标签公式：

$$y_i := \mathbb{E}_{a' \sim \pi}\left[r + \gamma \max_{a'} Q(s',a';\theta_{i-1})\right]$$

在该公式中可以看到，总是选择当时的最大已知值，这诱使神经网络学习的值高得脱离实际。特别是函数近似估计更会导致过高估算。高估(可能会在本示例中发生)会导致策略不佳，并容易引起模型偏差。正如在《毁灭战士》游戏示例中所体现的那样，存在这样一个事实例证：agent 经常感到被迫射击，而不管其相对于敌人的位置如何。如何精确解决该问题？

3.9 双 Q 学习和双深度 Q 网络

如前面的公式中所强调的，max 运算符使用相同的值来选择和评估给定环境状态下的动作。准确地说，当将该过程分为两个独立的过程(选择过程和评估过程)时，便获得了双 Q 学习(Double Q Learning)。双 Q 学习使用两个值函数，每个函数都有两个独立的权重集。其中一个权重集用于确定 greedy-epsilon 的利用或探索的权衡问题，而另一个则用于确定给定动作的值。然后，目标可粗略重写为以下形式。

$$Y_t^Q = R_{t+1} + \gamma Q\left(S_{t+1}, \arg\max_a Q(S_{t+1},a;\theta_t);\theta_t\right)$$

解释完这一点之后，现在可以讨论双 Q 网络(Double Q Network)以及如何利用它来克服深度 Q 网络的不足。我们没有添加其他模型，而是利用目标网络来评估值，同时利用在线网络来评估探索-利用决策过程。双 Q 网络的目标函数如下：

$$Y_t^{\text{Double QN}} \equiv R_{t+1} + \gamma Q\left(S_{t+1}, \arg\max_a Q(S_{t+1},a;\theta_t),\theta_t^-\right)$$

3.10 本章小结

在完成 Q 学习和深度 Q 学习两个示例的学习之后,建议读者尝试在不同背景下应用这些算法。必要时,可以更改参数并派生 fork/更改现有代码和模型。无论如何,建议读者在阅读后续章节时记住以下几点。

- Q 学习简单明了,易于解释——该算法的优点在于,易于理解 Q 值的输入原理。对于那些要求算法看上去一目了然的任务,考虑应用该算法实乃明智之举。
- 对于大型状态空间,Q 学习具有局限性——尽管前面的介绍只针对简单问题,但重要的是需要认识到,在类似《毁灭战士》游戏示例和更复杂的环境中,普通 Q 学习算法将花费大量时间才能完成训练。
- 深度 Q 学习仍然可能会陷入局部最优——与其他强化学习算法一样,DQN 仍然可能找到局部最优策略,而不是全局最优策略。从训练角度来看,寻找全局最优值的过程可能会非常耗时耗力。
- 尝试实现双 Q 学习和双深度 Q 网络——通过应用更先进的技术可以快速克服 Q 学习和 DQN 的局限性。从该点出发,读者应尝试从零开始实现最先进的算法。

学习完这些示例后,让我们继续讨论其他尚未涉及的强化学习算法,并深入讨论这些算法。

第 4 章
基于强化学习的做市策略

与只解决强化学习领域的一些标准问题(例如在许多书籍中出现的示例)不同的是,最好还能求解那些答案既不那么客观也未完全解决的领域。对于强化学习来说,做市(market making)是金融领域中最好的例子之一。我们将概述该学科(做市)的基础情况,介绍一些不基于机器学习的基本方法,然后测试几种基于强化学习的方法。

4.1 什么是做市

在金融市场中,进行交易的从业人员之间需要持续的流动性。在任何给定时刻,每个试图出售资产订单的人都不可能与想要购买的人完全匹配。这样,做市商在方便订单[这些订单源自那些通常想要通过金融手段(短期或长期)参与的人员,他们持有的时间长短各不相同]执行方面起着至关重要的作用。通常,做市被描述为金融市场从业人员可以在金融市场中持续赚钱的几种方式之一,做市与博彩模型相反,博彩模型基于较高风险的赌博,但可有条件地获得更高收益。接下来,我们尝试并了解做市使用的数据是什么,以及可以期待的结果是什么。图 4-1 是具有相关订单的订单簿(order book)示例。

Orderbook (XBTUSD)		
Price	Size	Total
11169.0	10,412	206,032
11168.0	47,329	195,620
11167.5	57,430	148,291
11167.0	9,999	90,861
11164.5	100	80,862
11164.0	240	80,762
11163.5	2,000	80,522
11163.0	32,370	78,522
11162.5	2,414	46,152
11162.0	43,738	43,738
11162.0 ↑		
11129.25 / 11129.79		
11161.5	161,603	161,603
11160.5	40,029	201,632
11160.0	388	202,020
11159.5	737	202,757
11159.0	8,400	211,157
11158.0	10,636	221,793
11157.5	21,100	242,893
11157.0	5,791	248,684
11156.5	38,563	287,247
11156.0	9,800	297,047

图 4-1　订单簿示例

图 4-1 是位于订单簿两侧的订单示例，代表了出价和要价。当某人将订单发送到交易所并使用限价订单时，他们欲出售的数量就位于订单簿上，直到订单成交为止。尽管交易所专用的成交算法可能会有所不同，但它们通常会尝试按照订单接收顺序完成订单，这使得最新的订单最后成交。利用限价单的好处在于，它们可以显著减少所谓的"市场影响力"。简单地说，每当有人要购买巨量商品

时，就会向市场发出巨量需求信号。这意味着我们可以利用自己的能力以获得最佳价格，并完成自己的订单。因此，交易者经常会打散他们的限价单，以尽可能地隐藏意图。

举一个更具体的做市示例，让我们想象一下，我们是一家拥有大量客户的金融交易中心，这些客户通常希望交易每一个大订单。但是，并非所有这些订单都是均匀分布的，使得每个想要购买的人都对应另一个想要出售的大客户。因此，我们决定通过提供非常优惠的费率来鼓励做市商，以提供流动性，使得这些大客户的订单便于成交。交易所越能吸引做市商，从而使得他们带来更多流动性，从而对于想要交易的人、特别是机构买家来说，该交易所越好。

做市的基本思想是，从业人员通常希望以任何给定的价格购买或出售一种金融产品，从而随着时间的推移，他们的策略会为他们带来回报。做市的主要吸引力在于，一旦确定了一种可扩展的成功策略，与对冲基金和其他交易机构可能采用的传统定向模型相比，该策略通常在相当长的时间内有效。除此之外，做市的风险较低。虽然这么说，从实际意义上讲，做市的主要困难是依赖于市场，根据市场的不同，可能需要大量资金来促进做市。根据该描述，我们将利用强化学习算法尝试开发一种较直观的策略制定方法，而不是尝试进行更传统的量化金融研究。

4.2 Trading Gym

类似于 OpenAI Gym，以及我们用来玩各种游戏(如《超级马里奥兄弟》和《毁灭战士》)的软件包的衍生产品，本章将使用 Trading Gym。这是一个开源项目，其目标是帮助用户在交易应用背景中轻松使用强化学习算法。在图 4-2 中，可以看到 Trading Gym 的可视化显示示例。

在 Trading Gym 环境中，读者通常可以使用三个选项。

(1) 购买金融产品
(2) 出售金融产品
(3) 保持当前头寸

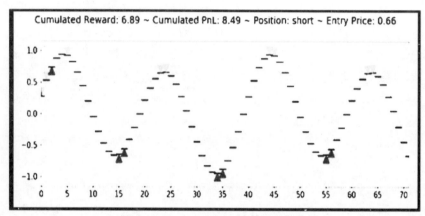

图 4-2　Trading Gym 的可视化显示

Trading Gym 通常允许用户处理一种(或多种)金融产品,其中数据的格式为(bid_product1, ask_product1, bid_product2, ask_product2)。我们将出价(bid)定义为个人可以购买产品的最佳可能价格,将要价(ask)定义为可以出售产品的最佳可能价格。我们将与读者一起逐步介绍如何将自己的订单数据导入该环境中,但在此之前,让我们首先讨论我们要解决的问题,并介绍解决该问题的较确定方法。

4.3　为什么强化学习适用于做市

尽管从 Trading Gym 并不容易看出来,但所有做市问题都需要使用限价单才能有效。市场订单的不足之处在于,因为总是要保证(在特定配额之下)成交订单所需的流动性,所以交易所通常收取可观的费用。因此,运用做市策略的唯一方法是在限价订单簿上下订

单并允许它们成交。根据该理由,这带来了一些问题。例如:

(1) 应该以什么价格买?

(2) 应该以什么价格卖?

(3) 应该以什么价格持有?

在机器学习领域,所有这些问题都不容易回答。具体来说,我们所处的空间是连续的。如前所述,市场在不断变化,对市场所采取的行动本身可能会导致订单更难执行或不执行。而普通的机器学习方法不会考虑这些环境因素,除非将环境因素作为特征包含进去。这样,除非事先对市场进行了大量研究,否则很难对其中的某些方面进行编码。大多数机器学习方法将是无效的,因为这是一个时间序列任务。唯一合适的方法是递归神经网络(RNN),特别是由于此类任务的粒度原因,将不得不预测相当大量的序列。这将导致形成一个模型,在该模型中,平均持仓时间比我们在做市过程中希望的时间长得多。我们需要敏捷性和灵活性,而使用机器学习方法可能会迫使我们在预定时间段内持有头寸,而不是根据市场环境在最有利情况下抛出头寸。所有这些原因证明了基于强化学习方法的合理性。下面开始描述代码,以及如何创建一个合理示例。以下是执行该功能的代码示例。

```
memory = Memory(max_size=memory_size)
environment = SpreadTrading(spread_coefficients=[1],
                            data_generator=generator,
                            trading_fee=trading_fee,
                            time_fee=time_fee,
history_length=history_length)
state_size = len(environment.reset())
```

在继续分析之前,首先应该了解一下 SpreadTrading()类中定义的一些重要属性。其中一些相当简单,由于金融市场上的所有交易通常都需要费用才能在交易所成交,因此必须设定一个交易费用。

在我们的第一个示例中,交换数据是合成的,第二个示例将使用真实订单数据。在示例中,将象征性收取不对应于任何真实交易所的交易费用。我们将 time_fee 设置为 0,因为应该没有该费用。但是,最重要的是,应该仔细讨论 DataGenerator 类及其作用。

4.4 使用 Trading Gym 合成订单簿数据

在使用 Trading Gym 时,可以选择直接处理订单簿数据或合成我们自己的数据。首先,将使用 WavySignal 函数,如下所示。

```
class WavySignal(DataGenerator):
    def _generator(period_1, period_2, epsilon, ba_spread=0):
        i = 0
        while True:
            i += 1
            bid_price = (1 - epsilon) * np.sin(2 * i * np.pi /
            period_1) + \
                epsilon * np.sin(2 * i * np.pi / period_2)
            yield bid_price, bid_price + ba_spread
```

为了让那些不熟悉 generator 函数的读者了解其含义,generator 函数通常用于需要迭代访问大量数据的实例,且我们已预先确定了应从何处读取这些数据;但是,鉴于正在解决应用的特性,这些数据实在太大,无法存储在内存中。取而代之的是存储对象。继续分析,该 generator(生成器)将根据前面的逻辑生成伪数据(fake data)。在 generator 正常工作的情况下,我们使用以下命令运行文件。

"pythonw -m chapter4.market_making_example"

在图 4-3 中我们观察一个类似的输出。

```
('Episode: 0', 'Total reward: -12.6', 'Loss: 2.69732347569e-10', 'Explore P: 0.9901')
('Episode: 1', 'Total reward: -13.755527185', 'Loss: 0.000434227811638', 'Explore P: 0.9804')
('Episode: 2', 'Total reward: -13.8', 'Loss: 0.00019776969566', 'Explore P: 0.9707')
('Episode: 3', 'Total reward: -13.0', 'Loss: 3.77897376893e-05', 'Explore P: 0.9612')
('Episode: 4', 'Total reward: -11.2', 'Loss: 0.000217975015403', 'Explore P: 0.9517')
```

图 4-3　WavySignal 数据生成器的输出

在本特定示例中，应用了一个深度 Q 网络来解决该问题。如我们所知，随着时间的推移，DQN 偏向于更多地利用环境，而不是强调探索。除了积累知识外，这使得我们取得比早期回合更高的分数。由于这是合成数据，因此没有继续分析此问题的必要理由。如果我们关注的重点是训练和选择算法，这将很有帮助。但是，在现实世界中，我们显然希望去解决问题，以便搞清楚在真实场景中可以采用的解决方案。

4.5 使用 Trading Gym 生成订单簿数据

在该环境中，我们有两种选择：(1)使用伪数据；(2)使用真实市场数据。除了可帮助熟悉环境的工作原理之外，我认为伪数据没有太大用处。因此，我们将开始利用真实数据。这将我们带到 CSVStreamer()类，如下所示。

```
class CSVStreamer(DataGenerator):
def _generator(filename, header=False):
        with open(filename, "r") as csvfile:
            reader = csv.reader(csvfile)
            if header:
            next(reader, None)
        for row in reader:
            #assert len(row) % 2 == 0
            yield np.array(row, dtype=np.float)
```

(代码经过节选，具体请参考 GitHub!)

CSVStreamer()类实质上可以通过_generator()函数进行概况，该函数简单地遍历文件中的每一行，且假设第一列是出价，第二列是要价。读者可以从 LOBSTER 下载数据，LOBSTER 允许用户获取不同的订单簿数据，或寻求从诸如 Bloomberg 这样的提供商那里购买数据。可以通过以下 URL 访问 LOBSTER 资源库：https://lobsterdata.com/。

来自 Bloomberg 的数据显然是相当昂贵的，因此，它应该留给那些拥有大量研究预算的人员或在拥有 Bloomberg 终端的机构内工作的人员使用。在此示例中将使用的"生成器"变量是 CSVStreamer，它加载包含在该资源库中的订单簿数据。继续分析，让我们从了解该示例中将执行大部分计算的函数开始。

```
def train_model(model, environment):
(代码经过节选,具体请参考 GitHub!)
        while step < max_steps:

           step += 1; decay_step += 1

           action, explore_probability = exploit_explore(...)
           state, reward, done, info = environment.step(action))
```

与在第 3 章中展示的《毁灭战士》游戏示例相似，大多数代码最终都是同质且相似的。我们将使用与之前相同的方式来遍历环境，此外，这里将集中精力比较多种方法的性能并评估应该使用哪一种方法。

4.6 实验设计

尽管很少公开披露关于做市商使用什么算法的确切信息，但总的来说，我们希望利用一套简单的规则。以下算法将构成我们的对照组(control group)，并对为什么基于规则的系统优于随机生成选择集的系统提供参考。与其他实验一样，对照组的目的是，使得我们能够将我们模型的结果与对照组的进行比较，以查看是否超出了对照组设定的基准。新的方法集将组成实验组。我们将基于以下标准评估算法是否可行。

- 总体奖励
- 整个实验的平均奖励

事不宜迟，接着讨论如何组成对照组/基准算法。以下列出了我们的两种策略的要求。

第4章 基于强化学习的做市策略

策略 1(实验组)
- 随机选择所有选项。

策略 2(对照组)
- 随机选择购买、持有、出售。
- 如果头寸多头,则出售资产。
- 如果头寸做空,则购买资产。
- 如果我们持有现金头寸,则随机选择一个选项。

这两种策略的代码位于 baseline_model()函数中,如下所示。

```
def baseline_model(n_actions, info, random=True):
    if random == True:
        action = np.random.choice(range(n_actions),
        p=np.repeat(1/float(n_actions), 3))
        action = possible_actions[action]

    else:

        if len(info) == 0:
            action = np.random.choice(range(n_actions),
            p=np.repeat(1/float(n_actions), 3))
            action = possible_actions[action]

        elif info['action'] == 'sell':
            action = buy

        else:
            action = sell

    return action
```

读者现在应该熟悉"info(信息)"字典的概念,它用于显示环境中的相关信息。在 Trading Gym 中,信息词典用于显示最近的动作。如果我们持有现金,则字典将为空。如果有未结头寸,它将位于"action""buy"或"sell"键下。有时如果我们没有持有现金,则从最近一次动作中获利。对于上面提到的实验,我们将在 1000 次小实验(trial)中重复进行 100 次交易。最后,在训练了模型之后,我们将重复同样的方案并比较结果。分别使用两种策略，从实验中得出以下结果。

- Strategy 1 average reward - 30 890
- Strategy 2 average reward - 62 777

与这些实验相关的数据和分布如下(见图4-4和图4-5)。

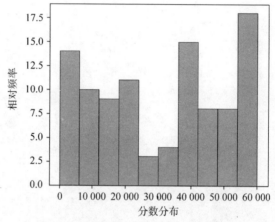

图4-4 基于随机选择动作的分数分布

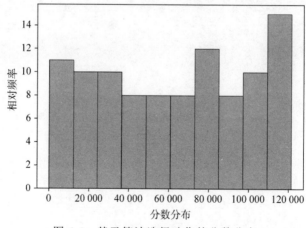

图4-5 基于算法选择动作的分数分布

第 4 章 基于强化学习的做市策略

如前面具有 1000 次小实验的实验所示，相比于在所有这些小实验过程中随机选择动作，当决策过程存在合理逻辑时，所选择动作的奖励结果要明显好得多。这样，如果根据该逻辑能够找到一个模型，该模型可最佳地选择出这些结果，而不是像之前那样简单地进行启发式搜索，则应该能进一步提高我们的收益。考虑清楚该方法后，尝试一下我们提出的解决方案。

4.6.1 强化学习方法 1：策略梯度

普通策略梯度方法固然有其不足，但其决策的数量相对有限，使我们可以轻松地迭代遍历各选项。该空间的缺点是我们可能捕获不了状态空间中的连续元素。综上所述，存在一个迫切需要解决的问题，即损失函数。当第一次使用策略梯度时，我们只有两个类，并且在离散样本空间中进行操作。因此，可以利用一个对数似然损失。但是，在本示例中，我们有多个类并且在连续空间中操作。这些是我们应该意识到的挑战，本书将在后面探讨其结果。

对于本示例，将使用分类交叉熵损失函数以及另一个自定义损失函数。前者是 Keras 的固有函数，通常用于包含两个以上类的分类方案中。

当运行前面设计的实验时，其结果非常之差。由于存在大量不同参数和不同样式，很大程度上不建议使用策略梯度法。考虑到这一点，接下来我们尝试深度 Q 网络。

4.6.2 强化学习方法 2：深度 Q 网络

在本示例中，就构思问题而言，Q 学习毫无疑问是一种绝佳选择，但深度 Q 学习应该是我们最终选择的方法。其背后的原因在于，状态空间(尤其是在考虑众多选项时)可能会非常大。运行函数的这一部分代码，应该得到类似于图 4-6 的输出。

```
('Episode: 0', 'Total reward: -133.4', 'Loss: 0.00609607854858', 'Explore P: 0.9058')
('Episode: 1', 'Total reward: -115.6', 'Loss: 0.00257436116226', 'Explore P: 0.8205')
('Episode: 2', 'Total reward: -102.4', 'Loss: 0.0003326706583681', 'Explore P: 0.7434')
('Episode: 3', 'Total reward: -98.0', 'Loss: 0.003310506671667', 'Explore P: 0.6736')
('Episode: 4', 'Total reward: 1383.0', 'Loss: 0.00124389817938', 'Explore P: 0.6105')
('Episode: 5', 'Total reward: -74.4', 'Loss: 0.000752292573452', 'Explore P: 0.5533')
('Episode: 6', 'Total reward: -63.4', 'Loss: 0.0001555577619', 'Explore P: 0.5016')
('Episode: 7', 'Total reward: -65.4', 'Loss: 0.000629861897323', 'Explore P: 0.4548')
('Episode: 8', 'Total reward: -58.4', 'Loss: 0.000218583256355', 'Explore P: 0.4125')
('Episode: 9', 'Total reward: -56.6', 'Loss: 0.000753238273319', 'Explore P: 0.3742')
('Episode: 10', 'Total reward: -46.2', 'Loss: 0.000115807146358', 'Explore P: 0.3395')
('Episode: 11', 'Total reward: -40.0', 'Loss: 0.000382943660952', 'Explore P: 0.3082')
('Episode: 12', 'Total reward: -35.0', 'Loss: 3.93268710468e-05', 'Explore P: 0.2798')
('Episode: 13', 'Total reward: -40.8', 'Loss: 0.000602947198786', 'Explore P: 0.2541')
('Episode: 14', 'Total reward: -31.0', 'Loss: 0.000622002757154', 'Explore P: 0.2309')
('Episode: 15', 'Total reward: -30.4', 'Loss: 4.48514838354e-05', 'Explore P: 0.2099')
('Episode: 16', 'Total reward: -24.8', 'Loss: 0.000435164460214', 'Explore P: 0.1909')
('Episode: 17', 'Total reward: -22.8', 'Loss: 0.000406970444601', 'Explore P: 0.1736')
('Episode: 18', 'Total reward: -24.6', 'Loss: 6.83790640323e-05', 'Explore P: 0.1581')
```

图 4-6　DQL 模型训练示例截图

在进行多次迭代训练期间,鉴于所拥有的数据量,根据观察结果,不建议进行一个回合以上的训练。尽管如此,有时取得的结果差异非常大。在某些迭代中,观察到的结果非常好,模型的某些结果根本不会选择任何动作,或选择某些动作。在某些情况下,观察到在此应用背景下做市算法在训练中确实表现出色,但结果不稳定或不一致。我注意到,绝大多数情况下,所建议的模型表现不佳的情况多于其他情况,并且经常做出不符合需要的决策。但是,继续前进,我们看一下重复进行样本外试验时的结果(见图 4-7)。

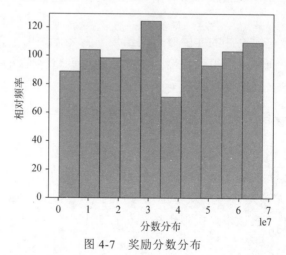

图 4-7　奖励分数分布

上述结果不仅明显优于基准算法，而且显著优于策略梯度模型，因此显而易见选择该模型。平均奖励为34、286、348，这绝对是一种可行的解决方案。如图4-7所示，我们的得分是令人满意的，并且似乎呈现双峰分布。

4.7 结果和讨论

在分析了所有结果之后，可以客观地宣布，读者既不应使用深度Q学习算法，也不应使用策略梯度法。总而言之，以下是我们这样建议的原因。

- 实验的算法超过了基准算法吗？为了证明任何实验方法的合理性，其结果必须超过基准算法。值得检查的是，数据采样方法是否合理，或者是否有足够的数据用于此特定实验。
- 某些算法赔钱了吗？对本章采用的第一种方法的最客观批评是，该方法没有达到其业务目标，即产生一种可盈利的策略。不建议在业务环境中应用该算法，并且是否应用最终也超出了理论研究的工作范畴，我们必须选择实际可行的算法。

继续前行，为了寻找可能有效的解决方案，可阅读此领域中的一些现有文献，以尝试对本章中的方案进行修订。尽管如此，许多公开发表的论文都遇到了这样的问题：算法暂时可盈利，但最终却无法获利。读者在自己的空闲时间中，还应该继续尝试使用ActorCritic方法，但也应该不惧怕尝试其他现有的解决方案，尝试此处未实验的不同的参数、费用结构和对策略的不同约束等。强化学习研究的难点在于，奖励函数设计是一个抽象过程，但它却被认为是优秀实验设计的关键组成部分。

4.8 本章小结

在完成以上示例后,我们到达本章末尾。在本章中从零开始解决了一些强化学习问题,并尝试改进了一些现有方法。本章主要内容是在尝试创建可部署解决方案时遇到的那些难点,本章也为读者建议了一个框架,展示了如何针对不同的样本成功获得较好的结果,这也表明了我们越来越接近最终答案。在此之前,处理的问题都相对简单明了或是一些经典示例,它们的价值在于能够"透明"地展示算法的功能。现在,我们终于到达本主题的难点部分,即学习如何推动各种解决方案的应用。

下面将进入最后一章,我们将重复此过程(利用特定算法解决特定问题的过程),但是在一个全新的环境中,我们将引导读者逐步了解如何从零开始创建自己的 OpenAI Gym 环境,以便他们能够自己开始研究!

第 5 章
自定义 OpenAI 强化学习环境

在本书的最后一章中，将重点介绍 OpenAI 的 gym 软件包，但更重要的是，尝试理解如何创建自己的自定义环境，以便解决不限于典型用例的各类问题。本章大部分内容将围绕我对 OpenAI 编程实践的建议，以及有关我通常如何编写大多数该类软件代码的建议。最后，在完成环境创建之后，将继续专注于解决问题。在本例中，将专注于尝试创建和解决一个新游戏。

5.1 《刺猬索尼克》游戏概述

《刺猬索尼克》(见图 5-1)是另一款经典游戏，通常被认为是《超级马里奥兄弟》的竞争对手。该游戏的概念是，在各级别关卡中，玩家操作刺猬从一侧赛跑到另一侧，目标是规避或杀死敌人并收集戒指。如果玩家受到攻击，他们将失去所有戒指。如果他们没有戒指时受到攻击，则会失去一次生命。如果他们失去全部生命，则游戏结束。目前我们不关注与 boss 战斗的任何级别的关卡，而是关注简单的入门级关卡(Level 1)。与这项任务相关的是，我们的目标是训练 agent 成功通过关卡而不死。

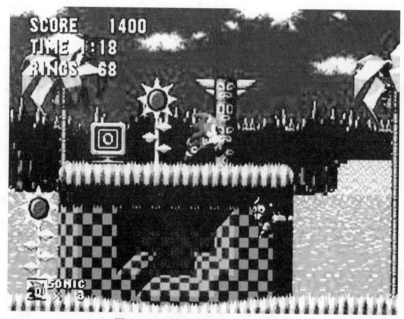

图 5-1 《刺猬索尼克》游戏截屏

5.2 下载该游戏

首先用户需要创建一个 Steam 账户，然后将 Steam 下载到本地计算机(如果尚未下载)。为了让那些不熟悉 Steam 的人了解情况，这里简单介绍一下。Steam 是一款游戏流服务，可以让玩家购买和租用游戏，而不必使用专门的控制台。在本应用背景中，我们将购买《刺猬索尼克》(*Sonic the Hedgehog*, $4.99)。用户下载完游戏，登录到 Steam 桌面客户端后，应该会看到图 5-2 所示的界面。

图 5-2　Steam 界面

安装该游戏后，读者应该能够看到 PLAY 按钮，表明已完成初步设置。但是，我们需要使用 retro 库来做一些样板类工作，现在将带领读者熟悉该库。retro 库专门用于处理版本较老的游戏，并完成这些游戏与 OpenAI 的兼容工作。这将解决许多原本需要我们去处理的繁重工作，从而加快处理过程。不考虑其他，让我们下载所需的文件。首先，通过以下 URL 下载并克隆 retro 库：https://github.com/openai/retro。

克隆完该库后，需要创建一个虚拟环境。为了让那些不熟悉的人了解情况，这里进行一下简介，虚拟环境是一种创建独立实例(即关于特定版本的 Python 及其相关依赖模块)的方法。这样做的好处是，对于独立的任务或项目，可以创建相应的 Python 环境，其中仅安装它们使用的依赖模块。安装了 virtualenv 之后，可以通过在 bash 终端中输入以下命令来实例化它。

```
"sudo mkdir virtual_environments && cd virtual_environments"
"virtualenv [environment name]/python3 -m venv [environment name]"
```

上述命令首先创建虚拟环境目录，然后利用 cd 命令进入该目

录,最后创建虚拟环境。完成此操作后,用户应使用 cd 命令进入本地克隆的 retro 库所在的目录。之后输入以下命令:

```
"python -m retro.import.sega_classics"
```

该命令将 sega_classics.py 文件中游戏的相应 ROM 写入本地环境。ROM 是指只读存储器,在本应用背景下通常是指保存了游戏的存储器,其中游戏常常是通过卡带(cartridge)发行的,这是磁盘和 DVD 出现之前的存储方式。现在,已经下载了该游戏及其相应的 ROM,接着进一步介绍如何使用 retro 和 Python 创建自定义环境。

5.3 编写该环境的代码

回顾《超级马里奥兄弟》和《毁灭战士》游戏示例,读者可以发现我们使用了一个自定义库,该库利用了一些相同技术。首先,我们分析 chapter5/create_environment.py 中的函数,并详细描述每个函数的功能。先看一下如下所示的函数:

```
def create_new_environment(environment_index, n_frames=4):
    (代码经过节选,具体请参考 GitHub!)
    print(dictionary[environment_index]['game'])

    print(dictionary[environment_index]['state'])

    environment = make(game=dictionary[environment_index]
    ['game'],

    state=dictionary[environment_index]['state'],
    bk2dir="./records")

    environment = ActionsDiscretizer(environment)
    environment = RewardScaler(environment)
    environment = PreprocessFrame(environment)
    environment = FrameStack(environment, n_frames)
    environment = AllowBacktracking(environment)
    return environment
```

第 5 章　自定义 OpenAI 强化学习环境

创建环境的过程非常简单，只不过是将参数从 retro_contest 模块传递给 make() 函数。这样就创建了一个环境，然后可以利用各种函数在其中添加结构，直到最终返回自定义和格式化的环境。接下来我们介绍一下环境的最重要方面之一，即创建和定义可以在环境中执行的动作。

```
class PreprocessFrame(gym.ObservationWrapper):
def __init__(self, environment, width, height):
    gym.ObservationWrapper.__init__(self, environment)
    self.width = width
    self.height = height
    self.observation_space = gym.spaces.Box(low=0,
high=255,
shape=(self.height, self.width, 1),
dtype=np.uint8)

    def observation(self, image):
        image = cv2.cvtColor(image, cv2.COLOR_RGB2GRAY)
        image = cv2.resize(image, (self.width, self.height),
interpolation=cv2.INTER_AREA)
        image = image[:, :, None]
        return image
```

像在处理 2D 或 3D 游戏时遇到的大多数问题一样，我们本质上是在处理计算机视觉问题中图像的一个排列。因此，需要先对图像进行预处理，以降低输入图像大小或将利用的神经网络(或其他方法)的复杂度。预处理后通常返回灰度图像的单个一维矩阵。对于阅读了前几章的读者，应该熟悉其中的大部分代码，我们首先实例化 PreprocessFrame() 类，该类首先将 ObservationWrapper 作为其唯一参数。从 OpenAI Gym 源代码可以看出，读者在前面的每个示例中都曾使用过该类，如下所示。

```
class ObservationWrapper(Wrapper):
    def reset(self, **kwargs):
        observation = self.env.reset(**kwargs)
        return self.observation(observation)

    def step(self, action):
```

```
            observation, reward, done, info = self.env.step(action)
            return self.observation(observation), reward, done, info
    def observation(self, observation):
        raise NotImplementedError
```

这是 retro 库的核心,可以在其中步进、重置并产生环境的当前状态。回到 PreprocessFrame()类,首先定义环境、要输出的图像的宽度和高度。根据这三个参数,还定义了观察空间,以便能够在其中操纵 agent。为此,这里应用了 Gym 的 Box()类。简单地将其定义为欧几里得空间 $\mathbb{R}^n$ 中的元素。在本例中,我们将此盒子(box)的边界定义为 0 和 255,代表给定像素的白度,其中 0 表示完全没有白色(黑色),而 255 表示完全没有黑色(白色)。observation()函数执行单帧的灰度化并输出它,以便对其进行分析。接着将深入探讨如何利用下一个类 ActionsDiscretizer()创建环境。

```
class ActionsDiscretizer(gym.ActionWrapper):
def __init__(self, env):
        super(ActionsDiscretizer, self).__init__(env)
        buttons = ["B", "A", "MODE", "START", "UP", "DOWN",
        "LEFT", "RIGHT", "C", "Y", "X", "Z"]
        actions = [['LEFT'], ['RIGHT'], ['LEFT', 'DOWN'],
        ['RIGHT', 'DOWN'], ['DOWN'],
                    ['DOWN', 'B'], ['B']]
        self._actions = []
```

该类从类的实例化开始,读卡器应该指向 buttons 数组和 actions 数组。根据是为键盘还是为特定的游戏控制器设计环境,按钮将有所不同。示例中的这些按钮对应于 Sega Genesis 控制器上的所有可能按钮。

需要指出的是,并不是所有可能的动作都会映射到每个按钮,尤其是在此版本的《刺猬索尼克》游戏中。尽管在该游戏的较新版本中添加了某些高级功能,但初始版本的游戏还是很标准的,在该版本中 Sonic 可以左右移动或奔跑,并可以使用 B 按钮跳。接着看一下如何创建特定的动作空间。

第5章 自定义OpenAI强化学习环境

```
for action in actions:
    _actions = np.array([False] * len(buttons))
    for button in action:
        _actions[buttons.index(button)] = True
    self._actions.append(_actions)
self.action_space = gym.spaces.Discrete(len(self._actions))
```

对于 actions 数组,迭代遍历该数组中的每个动作,然后创建一个名为_actions 的新数组。该数组是一个大小为 1×N 的数组,其中 N 是控制器上的按钮数,每个索引处的值均为 false。现在,对于 actions 数组中的每个按钮,我们希望将其映射到一个数组,其中某些条目为 false,其他条目为 true。最后,将其赋值给 action_space,作为 self 变量的一个属性。我们已经多次讨论了奖励的标准化(或归一化),因此不必再次介绍该函数。但是,应该讨论一个重要函数,尤其是在与本例类似的游戏/环境中。

```
class AllowBacktracking(gym.Wrapper):
def __init__(self, environment):
    super(AllowBacktracking, self).__init__(environment)
    self.curent_reward = 0
    self.max_reward = 0

def reset(self, **kwargs):
    self.current_reward = 0
    self.max_reward = 0
    return self.env.reset(**kwargs)
```

AllowBacktracting()类非常简单,因为在 2D 环境中,必须通过向后移动以最终到达该级别关卡的末端。需要指出的是,有时候如果偶尔(概率很小)回退我们的步骤,然后选择另外一系列动作,则有可能存在一条更好的道路可走。但我们不想鼓励奖励机制过多,因此将以下步骤函数分配给环境。

```
def step(self, action):
    observation, reward, done, info = self.environment.
    step(action)
    self.current_reward += reward
    reward = max(0, self.current_reward - self.max_reward)
```

```
self.max_reward = max(self.max_reward, self.current_
reward)
return observation, reward, done, info
```

对于该函数，读者需要重点关注的是，我们将奖励赋值为 0 或大于 0 的值。在这种情况下，如果奖励值不佳，则不会回退。完成所有样板类(boilerplate)分析工作后，让我们继续讨论将具体使用的模型及这样选择的原因。

5.4　A3C Actor-Critic

读者应该记得，在尝试训练 agent 玩《超级马里奥兄弟》游戏时，使用了 Advantage Actor-Critic 模型(简称 A2C)。在图 5-3 中，可以看到 A3C 网络的可视化展示。

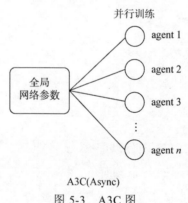

图 5-3　A3C 图

如前所述，Actor-Critic 网络能够使用值函数来更新策略函数，从该意义上来说它是有效的。不必等待执行所有动作到回合完成，也不管哪个动作是好是坏，我们可以逐步评估每个动作，然后相应改变策略，以得到更优结果，这比使用普通策略梯度的速度更快。与 A2C 相比，A3C 往往不太理想，因为我们基于一组初始全局参数来训练多个彼此平行的 agent。每个 agent 在探索环境时，将相应地

第 5 章 自定义 OpenAI 强化学习环境

更新参数,其他 agent 将根据这些参数更新。但是,并非所有 agent 都同时更新,因此该问题具有"异步"性质。接着继续讨论我们的实现,它包含在 A3CModel()类中。

```
class A3CNetwork():

    def __init__(self, s_size, a_size, scope, trainer):
        (code redacted)
layer3 = tf.layers.flatten(inputs=layer3)

        output_layer = fully_connected_layer(inputs=layer3,
        units=512,
activation='softmax')

        outputs, cell_state, hidden_state = lstm_
layer(input=hidden,
size=s_size,
actions=a_size,
apply_softmax=False)
```

与之前部署的 A2C 解决方案类似,首先传递预处理过的图像通过卷积层。如前所述,预处理有助于减少尺寸并消除数据中的噪声。但是,我们将在此处采用一个新步骤(不是前面示例中的步骤),该步骤将数据传递到 LSTM 层。LSTM 是 20 世纪 90 年代由 Sepp Hochreiter 和 Jürgen Schmidhuber 设计的模型——长短期记忆网络(LSTM)。首先可视化展示该模型的外观,如图 5-4 所示。

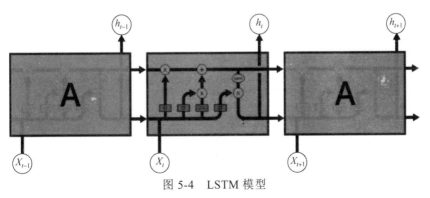

图 5-4 LSTM 模型

LSTM 在结构上的区别在于，可将它们看成块或单元，而不是通常看到的神经网络的传统结构。尽管如此，这里通常也适用相同的规则。我们对之前讨论的普通 RNN 的隐藏状态进行了改进，下面将开始逐步介绍与 LSTM 相关的公式。

$$i_t = \sigma\left(W_{xi}x_t + W_{hi}h_{t-1} + W_{hc}c_{t-1} + b_i\right)$$

$$f_t = \sigma\left(W_{xf}x_t + W_{hf}h_{t-1} + W_{hf}c_{t-1} + b_f\right)$$

$$c_t = f_t \circ c_{t-1} + i_t \circ \tanh\left(W_{xc}x_t + W_{hc}h_{t-1} + b_c\right)$$

$$o_t = \sigma\left(W_{xo}x_t + W_{ho}h_{t-1} + W_{co}c_t + b_o\right)$$

$$h_t = o_t \circ \tanh(c_t)$$

其中，i_t 是输入门，f_t 是遗忘门，c_t 是单元状态，o_t 是输出门，h_t 是输出向量，σ 是 Sigmoid 型激活函数，tanh 是 tanh 激活函数。

首先，关注一下该模型的图，特别是中心处的 LSTM 单元，并了解与公式相关的数据流。我们先讨论注释。每个块(用矩形表示)代表一个神经网络层，通过该层传递值。带箭头的水平线表示向量和数据移动方向。数据在移动通过神经网络层后，通常会传递给以圆表示的点操作对象。在算法初始化时，隐藏状态和单元状态都被初始化为 0。从编程方面来看，与 LSTM 层相关的大多数计算都在 TensorFlow 提供的 dynamic_rnn()函数内进行；但是，我们围绕该函数创建了一个主体函数，其中上述的单元、状态和关联变量的定义如下。

```
def lstm_layer(input, size, actions, apply_softmax=False):
    input = tf.expand_dims(input, [0])
    lstm = tf.contrib.rnn.BasicLSTMCell(size, state_is_
    tuple=True)
    state_size = lstm.state_size
    step_size = tf.shape(input)[:1]
    cell_init = np.zeros((1, state_size.c), np.float32)
```

第 5 章 自定义 OpenAI 强化学习环境

```
hidden_init = np.zeros((1, state_size.h), np.float32)
initial_state = [cell_init, hidden_init]
cell_state = tf.placeholder(tf.float32, [1, state_size.c])
hidden_state = tf.placeholder(tf.float32, [1, state_size.h])
input_state = tf.contrib.rnn.LSTMStateTuple(cell_state,
hidden_state)
```
(代码经过节选,具体请参考 GitHub!)

至于在哪里以及何时使用 LSTM,最常见的是将它们应用于基于序列的任务,在这些任务中给定输出取决于多个输入,这样的例子如拼写检查、语言翻译和预测时间序列等。与这项特定任务相关,我们正准备预处理数据,以便一次堆叠 4 帧。通常这样做是为了尝试模拟某种形式的运动,在此过程中,会根据之前的几个观察结果确定要采取的最佳可能动作。在本应用背景下,使用 RNN 算法的理由很简单。尽管 LSTM 不是必需的,但我们认为,向读者展示如何将更多和不同类型的机器学习模型结合在一起以解决本类问题非常有用。继续向前,我们将注意力转移回 A3C 网络本身,介绍该函数的后半部分。

```
self.policy = slim.fully_connected(output_layer,
a_size,
    activation_fn=tf.nn.softmax,
    weights_initializer=normalized_columns_
    initializer(0.01),
    biases_initializer=None)
self.value = slim.fully_connected(rnn_out, 1,
    activation_fn=None,
    weights_initializer=normalized_columns_
    initializer(1.0),
    biases_initializer=None)
```

获得 LSTM 的输出后,将其传递到一个全连接层,这样就定义了策略和价值函数,后续将利用它们生成输出矩阵。读者应独自观察梯度的计算和参数的更新。而且,工作的异步特性正是模型差异化的来源。现在,将介绍代码的最后部分,该部分可称为 main/master 函数。

```
    def play_sonic()
(代码经过节选,具体请参考 GitHub!)
wiith tf.device("/cpu:0"):
        master_network = AC_Network(s_size,a_size,'global',None)
        num_workers = multiprocessing.cpu_count()
        workers = []

for i in range(num_workers):
            workers.append(Worker(environment=environment,
                          name=i,
                          s_size=s_size,
                          a_sizse=a_size,
                          trainer=trainer,
                          saver=saver,
                          model_path=model_path))
```

在下面的代码中,首先创建一个包含全局参数的主网络,然后根据可用的 CPU 创建一些 workers 线程。上述方法将确保不消耗过多的内存且不使程序崩溃。然后,对于要创建的每个 workers 线程,在实例化之后将它们附加到一个数组中。接下来的代码部分是计算执行的重要地方。

```
            coord = tf.train.Coordinator()
sess.run(tf.global_variables_initializer())
worker_threads = []

        for worker in workers:

            worker_work = lambda: worker.work(max_episode_
            length=max_episode_length,
            gamma=gamma,
            master_network=master_network,
            sess=sess,
            coord=coord)
            _thread = threading.Thread(target=(worker_work))
            _thread.start()
            worker_threads.append(_thread)
        coord.join(worker_threads)
```

读者应该首先注意将要使用的 tf.train.Coordinator()函数以及线程库。对于 A3C 的实现,重要的是要了解我们在后端所做的以消除任

第 5 章 自定义 OpenAI 强化学习环境

何潜在混乱的工作。为了便于不了解线程的人学习，这里简介一下，线程是独立的执行流，因此多线程将允许在不同的处理器上运行进程。我们利用_thread 变量代表创建的线程，创建方式是向 threading.Thread 传递一个函数参数，在本例中为 worker_work 变量。该变量是由 worker.work()函数创建的，其主要代码如下。

```
def work(self,max_episode_length,gamma,sess,coord,saver):
    (代码经过节选,具体请参考 GitHub!)
while self.env.is_episode_finished() == False:
action_dist, value_function ,rnn_state = sess.run([self.local_
AC.policy,
            self.local_AC.value,
self.local_AC.state_out]...)}

                    action = np.random.choice(action_dist[0],
                    p=action_dist[0])
                    action = np.argmax(action_dist == action)

                    reward = self.env.make_action(self.
                    actions[action]) / 100.0
                    done = self.env.is_episode_finished()
episode_buffer.append([prior_state, action, reward, current_
state, done, value[0,0]])
                    episode_values.append(value[0,0])
```

首先通过执行计算图/A3C 模型来实例化一些变量。具体来说，在该代码的前面部分中，我们希望从产生的分布中随机选择动作。从这一点来看，所有内容似乎与前面的示例类似。接着在该环境中执行了一个动作，这样应该会产生某些奖励以及一个值函数。但是，对读者来说，新的内容是如何更新 workers 的 master 参数。这本身又与多线程示例相关联，特别是与 coord.join()函数相关联。在了解了线程以及将其与 A3C 实现联系起来的基础上，终于可以讨论之前提到的 tf.train.Coodinator()函数。该函数可用于协调所有线程终止时的行为，这是利用 join()函数完成的，该函数用于处理一个线程等待另一个线程完成时的情形。join()函数将导致主线程暂停并等待另一个线程完成。这正是 A3C 的异步特性在此问题中得以体现的地方！

5.5 本章小结

在训练了模型 10 小时后，我们观察到了合格的性能。但是，在关卡的某些地方经常存在一个问题，索尼克经常需要奔跑一个圆形路径。因此，建议进行更多的训练。尽管如此，我们已经初步注意到 agent 具有击败或避开敌人的能力，以及在收集金币的同时通过关卡的能力。虽然如此，但这阐明了该问题的难度。

读者必须意识到强化学习的难度所在。尽管强化学习仍然是一个广泛研究的领域，但是对于那些想要部署强化学习解决方案的人员，应该意识到，从零开始编写模型的代码是非常困难的，特别是 Actor-Critic 模型。甚至在没有介绍模型本身的情况下，我们花费了大量时间来构建用于处理环境的样板类代码。尽管这是一款简单的 2D 游戏，但存在相当复杂的环境，值得对其进行研究，而工程师则可完全专注于构建用于渲染和封装环境的工具。

至于解决该问题本身，在不对问题处理方式进行准确探讨的情况下，直接训练将会导致大量的时间浪费。合理构思分析问题并准备好尝试许多不同的方法，但是相对于其他方法，请在强化学习方法上花费更多的时间。此外，在设计环境时，请考虑可以使用什么样的奖励机制。例如，在《刺猬索尼克》游戏示例中，是希望优先捡更多的戒指，还是优先消灭敌人获得积分？显然，如果索尼克死亡，则应该产生最大负奖励。但是，对于索尼克死于从地图上摔下来，或死于被随机敌人杀死，则以你的判断，哪种情况更糟？所有这些考虑都会影响训练过程，但这些考虑应该是在初始时就应解决的高层次问题。

第 5 章　自定义 OpenAI 强化学习环境

我们鼓励读者利用这些示例中提供的代码,并在认为合适的地方对它们进行改进,例如,通过改进某些实现使得在计算上更加高效,也可以是提出更好的解决方案。通过协作,我们可以解决难度无与伦比的问题,共同推动这一领域的发展。

附录 A
源代码

本附录提供了本书初始版本的源代码。偶尔必要时会对该代码库进行更新,关于更新信息请查看 www.apress.com/9781484251263 上的 Github。

A.1 做市模型的程序

```
from collections import deque
class Memory():
    def __init__(self, max_size):
        self.buffer = deque(maxlen = max_size)

    def add(self, experience):
        self.buffer.append(experience)

    def sample(self, batch_size):
        buffer_size = len(self.buffer)
        index = np.random.choice(np.arange(buffer_size),
                            size=batch_size,
                            replace=True)
```

```python
        return [self.buffer[i] for i in index]

class DeepQNetworkMM():
def __init__(self, n_units, n_classes, state_size, action_
size, learning_rate):
    self.state_size = state_size
    self.action_size = action_size
    self.learning_rate = learning_rate
    self.n_units = n_units
    self.n_classes = n_classes

    self.input_matrix = tf.placeholder(tf.float32,
      [None, state_size])
    self.actions = tf.placeholder(tf.float32,
      [None, n_classes])
    self.target_Q = tf.placeholder(tf.float32, [None])

    self.layer1 = fully_connected_layer(inputs=self.input_
    matrix, units=self.n_units, activation='selu')

    self.hidden_layer = fully_connected_layer(inputs=self.
    layer1, units=self.n_units, activation='selu')

    self.output_layer = fully_connected_layer(inputs=self.
    hidden_layer, units=n_classes, activation=None)

    self.predicted_Q = tf.reduce_sum(tf.multiply(self.
    output_layer, self.actions), axis=1)

    self.error_rate = tf.reduce_mean(tf.square(self.
    target_Q - self.predicted_Q))

    self.optimizer = tf.train.RMSPropOptimizer(self.
    learning_rate).minimize(self.error_rate)
```

A.2 策略梯度的程序

```python
import keras.layers as layers
from keras import backend
from keras.models import Model
from keras.optimizers import Adam
from keras.initializers import glorot_uniform
class PolicyGradient():

    def __init__(self, n_units, n_layers, n_columns, n_outputs,
```

附录 A 源代码

```python
          learning_rate, hidden_activation, output_activation, loss_
function):
    self.n_units = n_units
    self.n_layers = n_layers
    self.n_columns = n_columns
    self.n_outputs = n_outputs
    self.hidden_activation = hidden_activation
    self.output_activation = output_activation
    self.learning_rate = learning_rate
    self.loss_function = loss_function

def create_policy_model(self, input_shape):
    input_layer = layers.Input(shape=input_shape)
    advantages = layers.Input(shape=[1])

    hidden_layer = layers.Dense(units=self.n_units,
    activation=self.hidden_activation, use_bias=False,
    kernel_initializer=glorot_uniform(seed=42))(input_
    layer)

    output_layer = layers.Dense(units=self.n_outputs,
    activation=self.output_activation, use_bias=False,
    kernel_initializer=glorot_uniform(seed=42))(hidden_
    layer)

def log_likelihood_loss(actual_labels, predicted_labels):
    log_likelihood = backend.log(actual_labels *
    (actual_labels - predicted_labels) + (1 - actual_
    labels) * (actual_labels + predicted_labels))
    return backend.mean(log_likelihood * advantages,
    keepdims=True)

if self.loss_function == 'log_likelihood':
    self.loss_function = log_likelihood_loss
else:
    self.loss_function = 'categorical_crossentropy'

policy_model = Model(inputs=[input_layer, advantages],
outputs=output_layer)
policy_model.compile(loss=self.loss_function,
optimizer=Adam(self.learning_rate))
model_prediction = Model(input=[input_layer],
outputs=output_layer)
return policy_model, model_prediction
```

A.3 模型

```python
import tensorflow as tf, numpy as np
from baselines.common.distributions import make_pdtype

activation_dictionary = {'elu': tf.nn.elu,
                         'relu': tf.nn.relu,
                         'selu': tf.nn.selu,
                         'sigmoid': tf.nn.sigmoid,
                         'softmax': tf.nn.softmax,
                         None: None}

def normalized_columns_initializer(standard_deviation=1.0):
  def initializer(shape, dtype=None, partition_info=None):
    output = np.random.randn(*shape).astype(np.float32)
    output *= standard_deviation/float(np.sqrt(np.
    square(output).sum(axis=0, keepdims=True)))
    return tf.constant(output)
  return initializer

def linear_operation(x, size, name, initializer=None, bias_init=0):
  with tf.variable_scope(name):
    weights = tf.get_variable("w", [x.get_shape()[1], size], initializer=initializer)
    biases = tf.get_variable("b", [size], initializer=tf.constant_initializer(bias_init))
    return tf.matmul(x, weights) + biases
def convolution_layer(inputs, dimensions, filters, kernel_size, strides, gain=np.sqrt(2), activation='relu'):

    if dimensions == 3:

        return tf.layers.conv1d(inputs=inputs,
                                filters=filters,
                                kernel_size=kernel_size,
                                kernel_initializer=tf.orthogonal_initializer(gain),
                                strides=(strides),
                                activation=activation_dictionary[activation])

    elif dimensions == 4:

        return tf.layers.conv2d(inputs=inputs,
```

```python
                         filters=filters,
                         kernel_size=kernel_size,
                         kernel_initializer=tf.
                         orthogonal_initializer(gain),
                         strides=(strides),
                         activation=activation_
                         dictionary[activation])

def fully_connected_layer(inputs, units, activation, gain=np.
sqrt(2)):
    return tf.layers.dense(inputs=inputs,
                         units=units,
                         activation=activation_
                         dictionary[activation],
                         kernel_initializer=tf.orthogonal_
                         initializer(gain))

def lstm_layer(input, size, actions, apply_softmax=False):
    input = tf.expand_dims(input, [0])
    lstm = tf.contrib.rnn.BasicLSTMCell(size, state_is_
    tuple=True)
    state_size = lstm.state_size
    step_size = tf.shape(input)[:1]
    cell_init = np.zeros((1, state_size.c), np.float32)
    hidden_init = np.zeros((1, state_size.h), np.float32)
    initial_state = [cell_init, hidden_init]
    cell_state = tf.placeholder(tf.float32, [1, state_size.c])
    hidden_state = tf.placeholder(tf.float32, [1, state_size.h])
    input_state = tf.contrib.rnn.LSTMStateTuple(cell_state,
    hidden_state)
    _outputs, states = tf.nn.dynamic_rnn(cell=lstm,
                                       inupts=input,
                                       initial_state=input_
                                       state,
                                       sequence_length=step_size,
                                       time_major=False)
    _cell_state, _hidden_state = states
    output = tf.reshape(_outputs, [-1, size])
    output_state = [_cell_state[:1, :], _hidden_state[:1, :]]
    output = linear_operation(output, actions, "logits",
    normalized_columns_initializer(0.01))
    output = tf.nn.softmax(output, dim=-1)
    return output, _cell_state, _hidden_state
def create_weights_biases(n_layers, n_units, n_columns, n_outputs):
    '''
    Creates dictionaries of variable length for differing
    neural network models
```

```
Arguments

n_layers - int - number of layers
n_units - int - number of neurons within each individual
layer
n_columns - int - number of columns within dataset

:return: dict (int), dict (int)
'''
weights, biases = {}, {}
for i in range(n_layers):
    if i == 0:
        weights['layer'+str(i)] = tf.Variable(tf.random_
        normal([n_columns, n_units]))
        biases['layer'+str(i)] = tf.Variable(tf.random_
        normal([n_columns]))
    elif i != 0 and i != n_layers-1:
        weights['layer'+str(i)] = tf.Variable(tf.random_
        normal([n_units, n_units]))
        biases['layer'+str(i)] = tf.Variable(tf.random_
        normal([n_units]))
    elif i != 0 and i == n_layers-1:
        weights['output_layer'] = tf.Variable(tf.random_
        normal([n_units, n_outputs]))
        biases['output_layer'] = tf.Variable(tf.random_
        normal([n_outputs]))

    return weights, biases
def create_input_output(input_dtype, output_dtype, n_columns,
n_outputs):
    '''
    Create placeholder variables for tensorflow graph

    '''
    X = tf.placeholder(shape=(None, n_columns), dtype=input_dtype)
    Y = tf.placeholder(shape=(None, n_outputs), dtype=output_dtype)
    return X, Y

class DeepQNetwork():

    def __init__(self, n_units, n_classes, n_filters, stride,
    kernel, state_size, action_size, learning_rate):
        self.state_size = state_size
        self.action_size = action_size
        self.learning_rate = learning_rate
        self.n_units = n_units
        self.n_classes = n_classes
```

附录 A 源代码

```python
self.n_filters = n_filters
self.stride = stride
self.kernel = kernel

self.input_matrix = tf.placeholder(tf.float32,
 [None, state_size])
self.actions = tf.placeholder(tf.float32,
 [None, n_classes])
self.target_Q = tf.placeholder(tf.float32, [None])

self.network1 = convolution_layer(inputs=self.input_
matrix,
                        filters=self.n_filters,
                        kernel_size=self.kernel,
                        strides=self.stride,
                        dimensions=4,
                        activation='elu')

self.network1 = tf.layers.batch_normalization(self.
network1, training=True, epsilon=1e-5)
self.network2 = convolution_layer(inputs=self.network1,
                        filters=self.n_filters*2,
                        kernel_size=int(self.
                        kernel/2),
                        strides=int(self.stride/2),
                        dimensions=4,
                        activation='elu')

self.network2 = tf.layers.batch_
normalization(inputs=self.network2, training=True,
epsilon=1e-5)

self.network3 = convolution_layer(inputs=self.network2,
                        filters=self.n_filters*4,
                        kernel_size=int(self.
                        kernel/2),
                        strides=int(self.
                        stride/2), dimensions=4,
                        activation='elu')

self.network3 = tf.layers.batch_
normalization(inputs=self.network3, training=True,
epsilon=1e-5)

self.network3 = tf.layers.flatten(inputs=self.network3)

self.output = fully_connected_layer(inputs=self.network3,
                        units=self.n_units,
```

```python
                                   activation='elu')
        self.output = fully_connected_layer(inputs=self.output,
                        units=n_classes, activation=None)

        self.predicted_Q = tf.reduce_sum(tf.multiply(self.
        output, self.actions), axis=1)

        self.error_rate = tf.reduce_mean(tf.square(self.
        target_Q - self.predicted_Q))

        self.optimizer = tf.train.RMSPropOptimizer(self.
        learning_rate).minimize(self.error_rate)

class ActorCriticModel():

    def __init__(self, session, environment, action_space,
    n_batches, n_steps, reuse=False):
    session.run(tf.global_variables_initializer())
    self.distribution_type = make_pdtype(action_space)
    height, weight, channel = environment.shape
    inputs_ = tf.placeholder(tf.float32, [height, weight,
    channel], name='inputs')
    scaled_images = tf.cast(inputs_, tf.float32)/float(255)

    with tf.variable_scope('model', reuse=reuse):
        layer1 = tf.layers.batch_normalization(convolution_
        layer(inputs=scaled_images,
      filters=32,
      kernel_size=8,
      strides=4,
      dimensions=3))

        layer2 = tf.layers.batch_normalization(convolution_
        layer(inputs=tf.nn.relu(layer1),
      filters=64,
      kernel_size=4,
      strides=2,
      dimensions=3))

        layer3 = tf.layers.batch_normalization(convolution_
    layer(inputs=tf.nn.relu(layer2),
      filters=64,
      kernel_size=3,
      strides=1,
      dimensions=3))

        layer3 = tf.layers.flatten(inputs=layer3)
```

```python
        output_layer = fully_connected_layer(inputs=layer3,
        units=512, activation='softmax')
        self.distribution, self.logits = self.distribution_
        type.pdfromlatent(output_layer, init_scale=0.01)
        value_function = fully_connected_layer(output_
        layer, units=1, activation=None)[:, 0]

    self.initial_state = None
    sampled_action = self.distribution.sample()

    def step(current_state, *_args, **_kwargs):
      action, value = session.run([sampled_action, value_
      function], {inputs_: current_state})
      return action, value

    def value(current_state, *_args, **_kwargs):
      return session.run(value_function, {inputs_:
      current_state})

    def select_action(current_state, *_args, **_kwargs):
      return session.run(sampled_action, {inputs_:
      current_state})
    self.inputs_ = inputs_
    self.value_function = value_function
    self.step = step
    self.value = value
    self.select_action = select_action
```

A.4 第 1 章

OpenAI 示例

```python
    import gym

    def cartpole():
        environment = gym.make('CartPole-v1')
        environment.reset()
        for _ in range(1000):
            environment.render()
            action = environment.action_space.sample()
            observation, reward, done, info = environment.
            step(action)
            print("Step {}:".format(_))
            print("action: {}".format(action))
```

```python
        print("observation: {}".format(observation))
        print("reward: {}".format(reward))
        print("done: {}".format(done))
        print("info: {}".format(info))

if __name__ == '__main__':
    cartpole()
```

A.5 第 2 章

车杆示例

```python
import gym, numpy as np, matplotlib.pyplot as plt
from neural_networks.policy_gradient_utilities import PolicyGradient

#Parameters
n_units = 5
gamma = .99
batch_size = 50
learning_rate = 1e-3
n_episodes = 10000
render = False
goal = 190
n_layers = 2
n_classes = 2
environment = gym.make('CartPole-v1')
environment_dimension = len(environment.reset())

def calculate_discounted_reward(reward, gamma=gamma):
    output = [reward[i] * gamma**i for i in range(0, len(reward))]
    return output[::-1]

def score_model(model, n_tests, render=render):
    scores = []
    for _ in range(n_tests):
        environment.reset()
        observation = environment.reset()
        reward_sum = 0
        while True:
            if render:
```

```python
        environment.render()
        state = np.reshape(observation, [1, environment_
        dimension])
        predict = model.predict([state])[0]
        action = np.argmax(predict)
        observation, reward, done, _ = environment.
        step(action)
        reward_sum += reward
        if done:
            break
    scores.append(reward_sum)

environment.close()
return np.mean(scores)
def cart_pole_game(environment, policy_model, model_
predictions):
    loss = []
    n_episode, reward_sum, score, episode_done = 0, 0, 0, False
    n_actions = environment.action_space.n
    observation = environment.reset()

    states = np.empty(0).reshape(0, environment_dimension)
    actions = np.empty(0).reshape(0, 1)
    rewards = np.empty(0).reshape(0, 1)
    discounted_rewards = np.empty(0).reshape(0, 1)

    while n_episode < n_episodes:

        state = np.reshape(observation, [1, environment_
        dimension])
        prediction = model_predictions.predict([state])[0]
        action = np.random.choice(range(environment.action_
        space.n), p=prediction)
        states = np.vstack([states, state])
        actions = np.vstack([actions, action])

        observation, reward, episode_done, info = environment.
        step(action)
        reward_sum += reward
        rewards = np.vstack([rewards, reward])
    if episode_done == True:

        discounted_reward = calculate_discounted_
        reward(rewards)
        discounted_rewards = np.vstack([discounted_rewards,
        discounted_reward])
```

```python
            rewards = np.empty(0).reshape(0, 1)

        if (n_episode + 1) % batch_size == 0:

            discounted_rewards -= discounted_rewards.mean()
            discounted_rewards /= discounted_rewards.std()
            discounted_rewards = discounted_rewards.squeeze()
            actions = actions.squeeze().astype(int)

            train_actions = np.zeros([len(actions), n_actions])
            train_actions[np.arange(len(actions)), actions] = 1

            error = policy_model.train_on_batch([states,
            discounted_rewards], train_actions)
            loss.append(error)

            states = np.empty(0).reshape(0, environment_
            dimension)
            actions = np.empty(0).reshape(0, 1)
            discounted_rewards = np.empty(0).reshape(0, 1)

            score = score_model(model=model_predictions,
            n_tests=10)

            print('''\nEpisode: %s \nAverage Reward: %s \
            nScore: %s \nError: %s'''
                    )%(n_episode+1, reward_sum/float(batch_
                    size), score, np.mean(loss[-batch_
                    size:]))
                if score >= goal:
                    break

            reward_sum = 0

        n_episode += 1
        observation = environment.reset()

    plt.title('Policy Gradient Error plot over %s Episodes'%(n_
    episode+1))
    plt.xlabel('N batches')
    plt.ylabel('Error Rate')
    plt.plot(loss)
    plt.show()

if __name__ == '__main__':

    mlp_model = PolicyGradient(n_units=n_units,
```

附录A　源代码

```
                        n_layers=n_layers,
                        n_columns=environment_dimension,
                        n_outputs=n_classes,
                        learning_rate=learning_rate,
                        hidden_activation='selu',
                        output_activation='softmax',
                        loss_function='log_likelihood')

    policy_model, model_predictions = mlp_model.create_policy_
    model(input_shape=(environment_dimension, ))

    policy_model.summary()

    cart_pole_game(environment=environment,
                   policy_model=policy_model,
                   model_predictions=model_predictions)
```

《超级马里奥》示例

```
import gym, numpy as np, matplotlib.pyplot as plt
from neural_networks.policy_gradient_utilities import
PolicyGradient

#Parameters
n_units = 5
gamma = .99
batch_size = 50
learning_rate = 1e-3
n_episodes = 10000
render = False
goal = 190
n_layers = 2
n_classes = 2
environment = gym.make('CartPole-v1')
environment_dimension = len(environment.reset())

def calculate_discounted_reward(reward, gamma=gamma):
    output = [reward[i] * gamma**i for i in range(0,
    len(reward))]
    return output[::-1]

def score_model(model, n_tests, render=render):
    scores = []
    for _ in range(n_tests):
        environment.reset()
        observation = environment.reset()
```

```python
        reward_sum = 0
        while True:
            if render:
                environment.render()
            state = np.reshape(observation, [1, environment_
            dimension])
            predict = model.predict([state])[0]
            action = np.argmax(predict)
            observation, reward, done, _ = environment.step(action)
            reward_sum += reward
            if done:
                break
            scores.append(reward_sum)

    environment.close()
    return np.mean(scores)

def cart_pole_game(environment, policy_model, model_
predictions):
    loss = []
    n_episode, reward_sum, score, episode_done = 0, 0, 0, False
    n_actions = environment.action_space.n
    observation = environment.reset()

    states = np.empty(0).reshape(0, environment_dimension)
    actions = np.empty(0).reshape(0, 1)
    rewards = np.empty(0).reshape(0, 1)
    discounted_rewards = np.empty(0).reshape(0, 1)

    while n_episode < n_episodes:

        state = np.reshape(observation, [1, environment_
        dimension])
        prediction = model_predictions.predict([state])[0]
        action = np.random.choice(range(environment.action_
        space.n), p=prediction)
        states = np.vstack([states, state])
        actions = np.vstack([actions, action])
    observation, reward, episode_done, info = environment.
    step(action)
    reward_sum += reward
    rewards = np.vstack([rewards, reward])

    if episode_done == True:

        discounted_reward = calculate_discounted_
```

```
        reward(rewards)
        discounted_rewards = np.vstack([discounted_rewards,
        discounted_reward])
        rewards = np.empty(0).reshape(0, 1)

        if (n_episode + 1) % batch_size == 0:

    discounted_rewards -= discounted_rewards.mean()
    discounted_rewards /= discounted_rewards.std()
    discounted_rewards = discounted_rewards.squeeze()
    actions = actions.squeeze().astype(int)

    train_actions = np.zeros([len(actions), n_actions])
    train_actions[np.arange(len(actions)), actions] = 1

        error = policy_model.train_on_batch([states,
        discounted_rewards], train_actions)
        loss.append(error)

        states = np.empty(0).reshape(0, environment_
        dimension)
        actions = np.empty(0).reshape(0, 1)
        discounted_rewards = np.empty(0).reshape(0, 1)

        score = score_model(model=model_predictions,
        n_tests=10)

            print('''\nEpisode: %s \nAverage Reward: %s \
            nScore: %s \nError: %s'''
                )%(n_episode+1, reward_sum/float(batch_
                size), score, np.mean(loss[-batch_size:]))

            if score >= goal:
                break

            reward_sum = 0

    n_episode += 1
    observation = environment.reset()
plt.title('Policy Gradient Error plot over %s Episodes'%(n_
episode+1))
plt.xlabel('N batches')
plt.ylabel('Error Rate')
plt.plot(loss)
```

```python
        plt.show()

if __name__ == '__main__':

    mlp_model = PolicyGradient(n_units=n_units,
                               n_layers=n_layers,
                               n_columns=environment_dimension,
                               n_outputs=n_classes,
                               learning_rate=learning_rate,
                               hidden_activation='selu',
                               output_activation='softmax',
                               loss_function='log_likelihood')

    policy_model, model_predictions = mlp_model.create_policy_
    model(input_shape=(environment_dimension, ))

    policy_model.summary()
    cart_pole_game(environment=environment,
                   policy_model=policy_model,
                   model_predictions=model_predictions)
```

A.6 第 3 章

《冰湖》示例

```
import os, time, gym, numpy as np

#Parameters
learning_rate = 1e-2
gamma = 0.96
epsilon = 0.9
n_episodes = 10000
max_steps = 100
environment = gym.make('FrozenLake-v0')
Q_matrix = np.zeros((environment.observation_space.n,
environment.action_space.n))

def choose_action(state):
    '''

    To be used after Q table has been updated, returns an action

    Parameters:
```

附录 A 源代码

```python
        state - int - the current state of the agent

    :return: int
    '''
    return np.argmax(Q_matrix[state, :])
def exploit_explore(prior_state, epsilon=epsilon, Q_matrix=Q_
matrix):
    '''
    One half of the exploit-explore paradigm that we will
    utilize

    Parameters

        prior_state - int - the prior state of the environment
        at a given iteration
        epsilon - float - parameter that we use to determine
        whether we will try a new or current best action

    :return: int
    '''

    if np.random.uniform(0, 1) < epsilon:
        return environment.action_space.sample()
    else:
        return np.argmax(Q_matrix[prior_state, :])

def update_q_matrix(prior_state, observation , reward, action):
    '''
    Algorithm that updates the values in the Q table to reflect
    knowledge acquired by the agent

    Parameters

        prior_state - int - the prior state of the environment
        before the current timestemp
        observation - int - the current state of the
        environment
        reward - int - the reward yielded from the environment
        after an action
        action - int - the action suggested by the epsilon
        greedy algorithm

    :return: None
    '''
```

```python
        prediction = Q_matrix[prior_state, action]
        actual_label = reward + gamma * np.max(Q_
        matrix[observation, :])
        Q_matrix[prior_state, action] = Q_matrix[prior_state,
        action] + learning_rate*(actual_label - prediction)

def populate_q_matrix(render=False, n_episodes=n_episodes):
    '''

    Directly implementing Q Learning (Greedy Epsilon) on the
    Frozen Lake Game
    This function populations the empty Q matrix
    Parameters

        prior_state - int - the prior state of the environment
        before the current timestemp
        observation - int - the current state of the environment
        reward - int - the reward yielded from the environment
        after an action
        action - int - the action suggested by the epsilon
        greedy algorithm

    :return: None
    '''

    for episode in range(n_episodes):
        prior_state = environment.reset()
        _ = 0
        while _ < max_steps:
            if render == True: environment.render()
            action = exploit_explore(prior_state)
            observation, reward, done, info = environment.
            step(action)

            update_q_matrix(prior_state=prior_state,
                            observation=observation,
                            reward=reward,
                            action=action)

            prior_state = observation
            _ += 1

            if done:
                break
def play_frozen_lake(n_episodes):
```

附录 A 源代码

```
'''
Directly implementing Q Learning (Greedy Epsilon) on the
Frozen Lake Game
This function uses the already populated Q Matrix and
displays the game being used

Parameters

    prior_state - int - the prior state of the environment
    before the current timestemp
    observation - int - the current state of the environment
    reward - int - the reward yielded from the environment
    after an action
    action - int - the action suggested by the epsilon
    greedy algorithm

:return: None
'''
for episode in range(n_episodes):
    print('Episode: %s'%episode+1)
    prior_state = environment.reset()
    done = False
    while not done:
        environment.render()
        action = choose_action(prior_state)
        observation, reward, done, info = environment.
        step(action)
        prior_state = observation
        if reward == 0:
            time.sleep(0.5)
        else:
            print('You have won on episode %s!'%(episode+1))
            time.sleep(5)
            os.system('clear')

        if done and reward == -1:
            print('You have lost this episode... :-/')
            time.sleep(5)
            os.system('clear')
            break

if __name__ == '__main__':

    populate_q_matrix(render=False)
    play_frozen_lake(n_episodes=10)
```

《毁灭战士》示例

```python
import warnings, random, time, tensorflow as tf, numpy as np, 
matplotlib.pyplot as plt
from neural_networks.models import DeepQNetwork
from algorithms.dql_utilities import create_environment, stack_
frames, Memory
from chapter3.frozen_lake_example import exploit_explore
from collections import deque

#Parameters
stack_size = 4
gamma = 0.95
memory_size = int(1e7)
train = True
episode_render = False
n_units = 500
n_classes = 3
learning_rate = 2e-4
stride = 4
kernel = 8
n_filters = 3
n_episodes = 1
max_steps = 100
batch_size = 64
environment, possible_actions = create_environment()
state_size = [84, 84, 4]
action_size = 3 #environment.get_avaiable_buttons_size()
explore_start = 1.0
explore_stop = 0.01
decay_rate = 1e-4
pretrain_length = batch_size
warnings.filterwarnings('ignore')
#writer = tf.summary.FileWriter("/tensorboard/dqn/1")
write_op = tf.summary.merge_all()

def exploit_explore(session, model, explore_start, explore_
stop, decay_rate, decay_step, state, actions):
    exp_exp_tradeoff = np.random.rand()
    explore_probability = explore_stop + (explore_start - 
    explore_stop) * np.exp(-decay_rate * decay_step)

    if (explore_probability > exp_exp_tradeoff):
        action = random.choice(possible_actions)

    else:
```

```python
        Qs = session.run(model.output, feed_dict = {model.
        input_matrix: state.reshape((1, *state.shape))})
        choice = np.argmax(Qs)
        action = possible_actions[int(choice)]

    return action, explore_probability
def train_model(model, environment):
    tf.summary.scalar('Loss', model.error_rate)
    saver = tf.train.Saver()
    stacked_frames = deque([np.zeros((84,84), dtype=np.int) for
    i in range(stack_size)], maxlen=4)
    memory = Memory(max_size=memory_size)
    scores = []

    with tf.Session() as sess:
        sess.run(tf.global_variables_initializer())
        decay_step = 0
        environment.init()

        for episode in range(n_episodes):
            step, reward_sum = 0, []
            environment.new_episode()
            state = environment.get_state().screen_buffer
            state, stacked_frames = stack_frames(stacked_
            frames, state, True)

            while step < max_steps:
                step += 1; decay_step += 1

            action, explore_probability = exploit_
            explore(session=sess,
                    model=model,
                    explore_start=explore_start,
                    explore_stop=explore_stop,
                    decay_rate=decay_rate,
                    decay_step=decay_step,
                    state=state,
                  actions=possible_actions)
                reward = environment.make_action(action)
                done = environment.is_episode_finished()
                reward_sum.append(reward)

                if done:

                    next_state = np.zeros((84,84), dtype=np.int)

                    next_state, stacked_frames = stack_
```

```python
            frames(stacked_frames=stacked_frames,
            state=next_state, new_episode=False)

            step = max_steps

            total_reward = np.sum(reward_sum)

            scores.append(total_reward)
            print('Episode: {}'.format(episode),
                  'Total reward: {}'.format(total_
                  reward),
                  'Explore P: {:.4f}'.
                  format(explore_probability))

            memory.add((state, action, reward, next_
            state, done))

        else:
            next_state = environment.get_state().
            screen_buffer
            next_state, stacked_frames = stack_
            frames(stacked_frames, next_state, False)
            memory.add((state, action, reward, next_
            state, done))
            state = next_state

        batch = memory.sample(batch_size)
        states = np.array([each[0] for each in batch],
        ndmin=3)
        actions = np.array([each[1] for each in batch])
        rewards = np.array([each[2] for each in batch])
        next_states = np.array([each[3] for each in
        batch], ndmin=3)
        dones = np.array([each[4] for each in batch])
        target_Qs_batch = []

        Qs_next_state = sess.run(model.predicted_Q,
        feed_dict={model.input_matrix: next_states,
        model.actions: actions})

        for i in range(0, len(batch)):
        terminal = dones[i]

        if terminal:
            target_Qs_batch.append(rewards[i])
        else:
            target = rewards[i] + gamma *
            np.max(Qs_next_state[i])
```

```python
                target_Qs_batch.append(target)

            targets = np.array([each for each in target_Qs_
            batch])

            error_rate, _ = sess.run([model.error_rate,
            model.optimizer], feed_dict={model.input_
            matrix: states, model.target_Q: targets, model.
            actions: actions})

            '''
            # Write TF Summaries
            summary = sess.run(write_op, feed_dict={model.
            inputs_: states, model.target_Q: targets,
            model.actions_: actions})

            writer.add_summary(summary, episode)
            writer.flush()

        if episode % 5 == 0:
            #saver.save(sess, filepath+'/models/model.ckpt')
            #print("Model Saved")
            '''

    plt.plot(scores)
    plt.title('DQN Performance During Training')
    plt.xlabel('N Episodes')
    plt.ylabel('Score Value')
    plt.show()
    plt.waitforbuttonpress()
    plt.close()
    return model

def play_doom(model, environment):

    stacked_frames = deque([np.zeros((84,84), dtype=np.int) for
    i in range(stack_size)], maxlen=4)
    scores = []

    with tf.Session() as sess:

        sess.run(tf.global_variables_initializer())
        totalScore = 0

        for _ in range(100):

            done = False
            environment.new_episode()
```

```python
            state = environment.get_state().screen_buffer
            state, stacked_frames = stack_frames(stacked_
            frames, state, True)

            while not environment.is_episode_finished():
                Q_matrix = sess.run(model.output, feed_dict =
                {model.input_matrix: state.reshape((1, *state.
                shape))})
                choice = np.argmax(Q_matrix)
                action = possible_actions[int(choice)]

                environment.make_action(action)
                done = environment.is_episode_finished()
                score = environment.get_total_reward()
                scores.append(score)
                time.sleep(0.01)

                if done:
                    break

            score = environment.get_total_reward()
            print("Score: ", score)

        environment.close()

    plt.plot(scores)
    plt.title('DQN Performance After Training')
    plt.xlabel('N Episodes')
    plt.ylabel('Score Value')
    plt.show()
    plt.waitforbuttonpress()
    plt.close()

if __name__ == '__main__':

    model = DeepQNetwork(n_units=n_units,
                        n_classes=n_classes,
                        n_filters=n_filters,
                        stride=stride,
                        kernel=kernel,
                        state_size=state_size,
                        action_size=action_size,
                        learning_rate=learning_rate)

    trained_model = train_model(model=model,
    environment=environment)

    play_doom(model=trained_model,
            environment=environment)
```

A.7 第 4 章

做市示例

```
import random, tensorflow as tf, numpy as np, matplotlib.pyplot as plt
from tgym.envs import SpreadTrading
from tgym.gens.deterministic import WavySignal
from neural_networks.market_making_models import DeepQNetworkMM, Memory
from chapter2.cart_pole_example import calculate_discounted_reward
from neural_networks.policy_gradient_utilities import PolicyGradient
from tgym.gens.csvstream import CSVStreamer

#Parameters
np.random.seed(2018)
n_episodes = 1
trading_fee = .2
time_fee = 0
history_length = 2
memory_size = 2000
gamma = 0.96
epsilon_min = 0.01
batch_size = 64
action_size = len(SpreadTrading._actions)
learning_rate = 1e-2
n_layers = 4
n_units = 500
n_classes = 3
goal = 190
max_steps = 1000
explore_start = 1.0
explore_stop = 0.01
decay_rate = 1e-4
_lambda = 0.95
value_coefficient = 0.5
entropy_coefficient = 0.01
max_grad_norm = 0.5
log_interval = 10
hold = np.array([1, 0, 0])
buy = np.array([0, 1, 0])
```

```python
sell = np.array([0, 0, 1])
possible_actions = [hold, buy, sell]

#Classes and variables
generator = CSVStreamer(filename='/Users/tawehbeysolow/
Downloads/amazon_order_book_data2.csv')
#generator = WavySignal(period_1=25, period_2=50, epsilon=-0.5)

memory = Memory(max_size=memory_size)

environment = SpreadTrading(spread_coefficients=[1],
                            data_generator=generator,
                            trading_fee=trading_fee,
                            time_fee=time_fee,
                            history_length=history_length)

state_size = len(environment.reset())

def baseline_model(n_actions, info, random=False):

    if random == True:
        action = np.random.choice(range(n_actions), p=np.
            repeat(1/float(n_actions), 3))
        action = possible_actions[action]

    else:

        if len(info) == 0:
            action = np.random.choice(range(n_actions), p=np.
                repeat(1/float(n_actions), 3))
            action = possible_actions[action]

        elif info['action'] == 'sell':
            action = buy

        else:
            action = sell

    return action

def score_model(model, n_tests):
    scores = []
    for _ in range(n_tests):
        environment.reset()
        observation = environment.reset()
        reward_sum = 0
        while True:
            "
```

附录 A 源代码

```python
        #environment.render()

        predict = model.predict([observation.reshape(1, 8)])[0]
        action = possible_actions[np.argmax(predict)]
        observation, reward, done, _ = environment.step(action)
        reward_sum += reward
        if done:
            break
    scores.append(reward_sum)

return np.mean(scores)
def exploit_explore(session, model, explore_start, explore_
stop, decay_rate, decay_step, state, actions):
    exp_exp_tradeoff = np.random.rand()
    explore_probability = explore_stop + (explore_start -
explore_stop) * np.exp(-decay_rate * decay_step)

    if (explore_probability > exp_exp_tradeoff):
        action = random.choice(possible_actions)

    else:
        Qs = session.run(model.output_layer, feed_dict =
{model.input_matrix: state.reshape((1, 8))})
        choice = np.argmax(Qs)
        action = possible_actions[int(choice)]

    return action, explore_probability

def train_model(environment, dql=None, pg=None, baseline=None):
    scores = []
    done = False
    error_rate, step = 0, 0
    info = {}
    n_episode, reward_sum, score, episode_done = 0, 0, 0, False
    n_actions = len(SpreadTrading._actions)
    observation = environment.reset()
    states = np.empty(0).reshape(0, state_size)
    actions = np.empty(0).reshape(0, len(
SpreadTrading._actions))
    rewards = np.empty(0).reshape(0, 1)
    discounted_rewards = np.empty(0).reshape(0, 1)
    observation = environment.reset()

    if baseline == True:

        for episode in range(n_episodes):
            for _ in range(100):
```

117

```python
            action = baseline_model(n_actions=n_actions,
                                info=info)

            state, reward, done, info = environment.step(action)
            reward_sum += reward

            next_state = np.zeros((state_size,), dtype=np.int)
            step = max_steps
            scores.append(reward_sum)
            memory.add((state, action, reward, next_state,
            done))

        print('Episode: {}'.format(episode),
            'Total reward: {}'.format(reward_sum))

        reward_sum = 0

    environment.reset()

    print(np.mean(scores))
    plt.hist(scores)
    plt.xlabel('Distribution of Scores')
    plt.ylabel('Relative Frequency')
    plt.show()
    plt.waitforbuttonpress()
    plt.close()

elif dql == True:

    loss = []

    model = DeepQNetworkMM(n_units=n_units,
                        n_classes=n_classes,
                        state_size=state_size,
                        action_size=action_size,
                        learning_rate=learning_rate)

    #tf.summary.scalar('Loss', model.error_rate)
    with tf.Session() as sess:

        sess.run(tf.global_variables_initializer())
        decay_step = 0

        for episode in range(n_episodes):

            current_step, reward_sum = 0, []
            state = np.reshape(observation, [1, state_size])
```

附录 A 源代码

```
    while current_step < max_steps:

        current_step += 1; decay_step += 1

        action, explore_probability = exploit_
        explore(session=sess,
model=model,
explore_start=explore_start,
explore_stop=explore_stop,
decay_rate=decay_rate,
decay_step=decay_step,
state=state,
actions=possible_actions)

            state, reward, done, info = environment.
            step(action)
            reward_sum.append(reward)

            if current_step >= max_steps:
                done = True

            if done == True:
            next_state = np.zeros((state_size,),
            dtype=np.int)
            step = max_steps
            total_reward = np.sum(reward_sum)
            scores.append(total_reward)
            memory.add((state, action, reward,
            next_state, done))

            print('Episode: {}'.format(episode),
                  'Total reward: {}'.
                  format(total_reward),
                  'Loss: {}'.format(error_rate),
                  'Explore P: {:.4f}'.
                  format(explore_probability))

            loss.append(error_rate)

        elif done != True:

            next_state = environment.reset()
            state = next_state
            memory.add((state, action, reward,
            next_state, done))

    batch = memory.sample(batch_size)
    states = np.array([each[0] for each in batch])
```

```python
        actions = np.array([each[1] for each in batch])
        rewards = np.array([each[2] for each in batch])
         next_states = np.array([each[3] for each in
         batch])
        dones = np.array([each[4] for each in batch])

        target_Qs_batch = []
        Qs_next_state = sess.run(model.predicted_Q,
        feed_dict={model.input_matrix: next_states,
        model.actions: actions})

        for i in range(0, len(batch)):
          terminal = dones[i]

          if terminal:
              target_Qs_batch.append(rewards[i])

          else:
              target = rewards[i] + gamma *
              np.max(Qs_next_state[i])
              target_Qs_batch.append(target)

        targets = np.array([each for each in
        target_Qs_batch])

        error_rate, _ = sess.run([model.error_rate,
        model.optimizer], feed_dict={model.input_
        matrix: states, model.target_Q: targets,
        model.actions: actions})
     if episode == n_episodes - 1:

        market_making(model=model,
                    environment=environment,
                    sess=sess,
                    state=state,
                    dpl=True)

  elif pg == True:

      loss = []
  mlp_model = PolicyGradient(n_units=n_units,
                            n_layers=n_layers,
                            n_columns=8,
                            n_outputs=n_classes,
                            learning_rate=learning_rate,
                            hidden_activation='selu',
                            output_activation='softmax',
```

附录A 源代码

```
                        loss_function='categorical_
                        crossentropy')
policy_model, model_predictions = mlp_model.create_
policy_model(input_shape=(len(observation),))

policy_model.summary()

while n_episode < n_episodes:

    state = observation.reshape(1, 8)
    prediction = model_predictions.predict([state])[0]
    action = np.random.choice(range(len(SpreadTrading._
    actions)), p=prediction)
    action = possible_actions[action]
    states = np.vstack([states, state])
    actions = np.vstack([actions, action])

    observation, reward, episode_done, info =
    environment.step(action)
    reward_sum += reward
    rewards = np.vstack([rewards, reward])
    step += 1

    if step == max_steps:
        episode_done = True
    if episode_done == True:

        discounted_reward = calculate_discounted_
        reward(rewards, gamma=gamma)
        discounted_rewards = np.vstack([discounted_
        rewards, discounted_reward])

        discounted_rewards -= discounted_rewards.mean()
        discounted_rewards /= discounted_rewards.std()
        discounted_rewards = discounted_rewards.squeeze()
        actions = actions.squeeze().astype(int)

        #train_actions = np.zeros([len(actions), n_
        actions])
        #train_actions[np.arange(len(actions)),
        actions] = 1

        error = policy_model.train_on_batch([states,
        discounted_rewards], actions)
        loss.append(error)
```

```python
            states = np.empty(0).reshape(0, 8)
            actions = np.empty(0).reshape(0, 3)
            rewards = np.empty(0).reshape(0, 1)
            discounted_rewards = np.empty(0).reshape(0, 1)

            score = score_model(model=model_predictions,
            n_tests=10)

            print("'\nEpisode: %s \nAverage Reward: %s \
            nScore: %s \nError: %s"'
                )%(n_episode+1, reward_sum/float(batch_
                size), score, np.mean(loss[-batch_
                size:]))
                if score >= goal:
                    break

            reward_sum = 0

            n_episode += 1
            observation = environment.reset()

        if n_episode == n_episodes - 1:

            market_making(model=model_predictions,
                    environment=environment,
                    sess=None,
                    state=state,
                    pg=True)
    if baseline != True:

        plt.title('Policy Gradient Error plot over %s
        Episodes'%(n_episode+1))
        plt.xlabel('N batches')
        plt.ylabel('Error Rate')
        plt.plot(loss)
        plt.show()
        plt.waitforbuttonpress()
        return model

def market_making(model, environment, sess, state, dpl=None,
pg=None):

    scores = []
    total_reward = 0
    environment.reset()

    for _ in range(1000):
        for __ in range(100):
```

```python
            state = np.reshape(state, [1, state_size])

            if dpl == True:
                Q_matrix = sess.run(model.output_layer, feed_dict
                = {model.input_matrix: state.reshape((1, 8))})
                choice = np.argmax(Q_matrix)
                action = possible_actions[int(choice)]

            elif pg == True:
                state = np.reshape(state, [1, 8])
                predict = model.predict([state])[0]
                action = np.argmax(predict)
                action = possible_actions[int(action)]

            state, reward, done, info = environment.step(action)
            total_reward += reward

            print('Episode: {}'.format(_),
                'Total reward: {}'.format(total_reward))
            scores.append(total_reward)
            state = environment.reset()

    print(np.mean(scores))
    plt.hist(scores)
    plt.xlabel('Distribution of Scores')
    plt.ylabel('Relative Frequency')
    plt.show()
    plt.waitforbuttonpress()
    plt.close()

if __name__ == '__main__':
    train_model(environment=environment, dql=True)
```

A.8 第 5 章

《刺猬索尼克》示例

```
import cv2, gym, numpy as np
from retro_contest.local import make
from retro import make as make_retro
from baselines.common.atari_wrappers import FrameStack

cv2.ocl.setUseOpenCL(False)
```

```python
class PreprocessFrame(gym.ObservationWrapper):
    """
    Grayscaling image from three dimensional RGB pixelated
    images
    - Set frame to gray
    - Resize the frame to 96x96x1
    """
    def __init__(self, environment, width, height):
        gym.ObservationWrapper.__init__(self, environment)
        self.width = width
        self.height = height
        self.observation_space = gym.spaces.Box(low=0,
                                  high=255,shape=(self.height,
                                  self.width, 1), dtype=np.
                                  uint8)

    def observation(self, image):
        image = cv2.cvtColor(image, cv2.COLOR_RGB2GRAY)
        image = cv2.resize(image, (self.width, self.height),
        interpolation=cv2.INTER_AREA)
        image = image[:, :, None]
        return image

class ActionsDiscretizer(gym.ActionWrapper):
    """
    Wrap a gym-retro environment and make it use discrete
    actions for the Sonic game.
    """
    def __init__(self, env):
        super(ActionsDiscretizer, self).__init__(env)
        buttons = ["B", "A", "MODE", "START", "UP", "DOWN",
        "LEFT", "RIGHT", "C", "Y", "X", "Z"]
        actions = [['LEFT'], ['RIGHT'], ['LEFT', 'DOWN'],
        ['RIGHT', 'DOWN'], ['DOWN'],
                   ['DOWN', 'B'], ['B']]
        self._actions = []
        """
        What we do in this loop:
        for each action in actions
            - Create an array of 12 False (12 = nb of buttons)
            for each button in action: (for instance ['LEFT'])
            we need to make that left button index = True
                - Then the button index = LEFT = True
            In fact at the end we will have an array where each
            array is an action and each elements True of this array
            are the buttons clicked.
        """
        for action in actions:
```

```python
            _actions = np.array([False] * len(buttons))
            for button in action:
                _actions[buttons.index(button)] = True
            self._actions.append(_actions)
        self.action_space = gym.spaces.Discrete(len(self._actions))
    def action(self, a):
        return self._actions[a].copy()

class RewardScaler(gym.RewardWrapper):
    """
    Bring rewards to a reasonable scale for PPO.
    This is incredibly important and effects performance
    drastically.
    """
    def reward(self, reward):

        return reward * 0.01

class AllowBacktracking(gym.Wrapper):
    """
    Use deltas in max(X) as the reward, rather than deltas
    in X. This way, agents are not discouraged too heavily
    from exploring backwards if there is no way to advance
    head-on in the level.
    """
    def __init__(self, environment):
        super(AllowBacktracking, self).__init__(environment)
        self.curent_reward = 0
        self.max_reward = 0

    def reset(self, **kwargs):
        self.current_reward = 0
        self.max_reward = 0
        return self.env.reset(**kwargs)

    def step(self, action):
        observation, reward, done, info = self.environment.
        step(action)
        self.current_reward += reward
        reward = max(0, self.current_reward - self.max_reward)
        self.max_reward = max(self.max_reward, self.current_
        reward)
        return observation, reward, done, info

def wrap_environment(environment, n_frames=4):
    environment = ActionsDiscretizer(environment)
    environment = RewardScaler(environment)
    environment = PreprocessFrame(environment)
    environment = FrameStack(environment, n_frames)
```

```python
    environment = AllowBacktracking(environment)
    return environment

def create_new_environment(environment_index, n_frames=4):
    """
    Create an environment with some standard wrappers.
    """
    dictionary = [
        {'game': 'SonicTheHedgehog-Genesis', 'state':
        'SpringYardZone.Act3'},
        {'game': 'SonicTheHedgehog-Genesis', 'state':
        'SpringYardZone.Act2'},
        {'game': 'SonicTheHedgehog-Genesis', 'state':
        'GreenHillZone.Act3'},
        {'game': 'SonicTheHedgehog-Genesis', 'state':
        'GreenHillZone.Act1'},
        {'game': 'SonicTheHedgehog-Genesis', 'state':
        'StarLightZone.Act2'},
        {'game': 'SonicTheHedgehog-Genesis', 'state':
        'StarLightZone.Act1'},
        {'game': 'SonicTheHedgehog-Genesis', 'state':
        'MarbleZone.Act2'},
        {'game': 'SonicTheHedgehog-Genesis', 'state':
        'MarbleZone.Act1'},
        {'game': 'SonicTheHedgehog-Genesis', 'state':
        'MarbleZone.Act3'},
        {'game': 'SonicTheHedgehog-Genesis', 'state':
        'ScrapBrainZone.Act2'},
        {'game': 'SonicTheHedgehog-Genesis', 'state':
        'LabyrinthZone.Act2'},
        {'game': 'SonicTheHedgehog-Genesis', 'state':
        'LabyrinthZone.Act1'},
        {'game': 'SonicTheHedgehog-Genesis', 'state':
        'LabyrinthZone.Act3'}]

    print(dictionary[environment_index]['game'])
    print(dictionary[environment_index]['state'])

    environment = make(game=dictionary[environment_index]['game'],
                state=dictionary[environment_index]['state'],
                bk2dir="./records")

    environment = wrap_environment(environment=environment,
                                n_frames=n_frames)

    return environment
def make_test_level_Green():
    return make_test()
```

附录 A 源代码

```python
def make_test(n_frames=4):
    """
    Create an environment with some standard wrappers.
    """
    environment = make_retro(game='SonicTheHedgehog-Genesis',
                             state='GreenHillZone.Act2',
                             record="./records")
    environment = wrap_environment(environment=environment,
                                   n_frames=n_frames)
    return environment
```